# The New
# Television Technologies

# The New Television Technologies

## Second Edition

Lynne Schafer Gross
California State University—Fullerton

wcb
Wm. C. Brown Publishers
Dubuque, Iowa

**To Kevin, Owen, and Brian to show
to their grandchildren**

# contents

**preface**  xiii

# part 1    introduction                                                    1

**Overview**

**1**

scope of the new television technologies                                    3
production technologies/ distribution technologies

interrelationships                                                          4
international implications/ new technology overlap/ traditional technologies/ regulation/ effects on the poor/ privacy, piracy, and pornography

organization of the book                                                    9
part 1/ part 2/ part 3/ part 4

**traditional distribution processes**

**2**

the electromagnetic spectrum                                               11
frequencies/ radio waves/ service placement

radio station broadcasting                                                 14
modulation/ am and fm characteristics/ bandwidth/ stereo/ multiplexing/ reception

television station broadcasting                                            17
stereo/ bandwidth/ spectrum placement/ multiplexing/ reception

wire transmission                                                          18
phone wires/ coaxial cable/ fiber optics

microwave                                                                  20
relay links/ uses

physical transportation                                                    20
old uses/ new uses

# part 2 foundations 23

**satellites**

**3**

description 25
positioning/ transmission/
transponders/ reception/
planned improvements

history 28
early communication satellites/
comsat and intelsat/ early
uses/ teleconferencing/ cable
tv use/ backyard reception/
additional uses

issues 32
international power/ backyard
satellite reception/ satellites
and the news/ teleconferencing
controversies

**computers**

**4**

description 37
basic operations/ processing/
memory/ input-output devices/
computer interconnections/
uses

history 42
the first generation/ the second
generation/ the third
generation/ the fourth
generation/ video games

issues 47
job replacement/ privacy/
decline of interpersonal
contact/ dependency/
economic downswings/
computer crime/ video games

# part 3 new technologies 53

**cable tv**

**5**

description 55
cable compared to broadcast/
headends, hubs, cable, and
converters/ cable
programming/ financial
considerations/ regulation

history 66
very early days of cable/
confused growth/ fcc actions/
early programming/ copyright
controversies/ the cable-
satellite alliance/ cable's gold
rush/ development of pay
cable/ basic cable growth/
changes in local programming/

interactive cable's success and stumbling/ consolidation and retrenchment/ the spirit of deregulation

**issues** 84
internal problems/ economic reality/ ownership/ relationship to government/ advertising effects on the poor/ national programming/ local programming/ interactive services/ piracy/ relationship to broadcasters

**subscription tv**

**6**

**description** 97
over-the-air scrambled transmission/ addressability/ single channel local operation/ programming/ sales/ regulation

**history** 102
early experiments/ bartlesville and etobicoke/ regulation/ hartford, california, and newark/ the peak of success/ piracy/ deregulation/ decline and fall

**issues** 106
competition/ cost/ movie depletion/ sexually explicit material/ local programming/ pay-per-view/ open airwaves concept

**low-power tv**

**7**

**description** 111
transmission characteristics/ regulation/ costs/ programming

**history** 113
translators/ popularity/ allocation of channels

**issues** 115
profitability/ ownership/ competition

**multichannel multipoint distribution service (mmds)**

**8**

**description** 117
broadcast and delivery/ distance/ ofs and itfs/ programming/ regulation

**history** 119
mds/ "wireless cable" concept/ itfs compromise/ lotteries/ mmds-itfs systems

issues     121
viability/ piracy/ educational
alliance/ cable conflicts/
visibility

## satellite master antenna TV (SMATV)

# 9

description     123
matv/ satellite dish addition/
regulation/ economics/
programming

history     125
early roots/ original builds/
cable competition/
congressional action

issues     126
cable conflicts/ piracy/ survival

## direct broadcast satellite

# 10

description     129
transmission and reception/
finances/ programming/
regulation

history     131
stc proposal/ other proposals/
high definition proposal/ interim
approval/ united satellite's
operation/ due diligence tests

issues     134
implementation/ relationship to
cable tv/ localism/
programming

## videocassettes

# 11

description     137
the video taping process/
formats/ features/ uses/
regulation

history     142
early recorders/ competition/
copyright suit/ rising
significance

issues     145
effects on broadcast tv/
relationship to ratings/ effects
on cable tv/ effects on the film
business/ program content/
technical standardization/
piracy

## videodiscs

# 12

description     151
formats/ programming/
regulation

history     155
early development/ initial
marketing/ a second try/
selectavision's demise/ laser
disc developments

|  |  | issues | 158 |
|  |  | technical options/ cost/ program innovation/ historical failure |  |

**teletext**

# 13

|  |  | description | 161 |
|  |  | over-the-air transmission/ reception/ cost/ programming/ graphics |  |
|  |  | history | 164 |
|  |  | the english system/ the french system/ the canadian system/ the beginnings of american teletext/ teletext experiments |  |
|  |  | issues | 168 |
|  |  | standardization/ technical problems/ programming limitations/ competition/ sociological effects |  |

**videotext**

# 14

|  |  | description | 171 |
|  |  | comparison to teletext/ two way wire delivery/ simple videotext/ videotext standards/ costs/ programming |  |
|  |  | history | 175 |
|  |  | the english system/ the french system/ the canadian system/ early text uses in the u.s./ videotext competition |  |
|  |  | issues | 181 |
|  |  | economic viability/ competition/ copyright, fraud, and obscenity/ privacy/ effects on the poor/ social ramifications |  |

**reception technologies**

# 15

|  |  | description | 187 |
|  |  | high definition tv/ digital tv/ flat screen tv/ large screen tv/ three-d tv |  |
|  |  | history | 192 |
|  |  | high definition tv/ digital tv/ flat screen tv/ large screen tv/ three-d tv |  |
|  |  | issues | 196 |
|  |  | high definition tv/ digital tv/ flat screen tv/ large screen tv/ three-d tv |  |

# part 4    conclusion                                     201
**implications and interrelationships**

# 16

interrelationships of media                               203
movies/ newspapers/
broadcasting/ cable tv/ other
new media

**social implications**                                   208
the individual/ society/
economics/ politics

**notes**   215
**glossary**   227
**index**   235

# preface

The first edition of this book was written at a time when the new television technologies seemed to be on a continuously upward spiral. Predictions were being made that traditional broadcasting would soon dwindle to a mere shell of itself and that consumers would be lured by the exciting, glamorous new media.

Such did not transpire. NBC, CBS, and ABC are still quite healthy and have actually changed little in the recent past. The new television technologies are the ones that have not fared so well. Indeed, as I was rewriting this book, I often felt as though I were writing obituaries.

And yet the future looks bright for the new technologies that are surviving. Perhaps the next edition of this book will be filled with reincarnations.

At any rate, this is an exciting field to write about and, I hope, an exciting field to read about. Because changes occur almost daily, this book can reflect only those made prior to its publication. I hope this edition is regarded as a snapshot in time that can be updated by faithful reading of the television trade magazines.

There are major differences between this book and the first edition. All the material has, of course, been updated to cover the three years since the first edition appeared. In addition, two totally new chapters have been included—one on traditional distribution processes and one on computers. The chapter on traditional distribution processes was included so that the reader could be brought up to date quickly on methods of distributing television programming material that existed before the new television technologies came to the fore. The chapter on computers is included because, although they are not strictly a television technology, computers play a crucial role in many of the TV technologies.

I am indebted to many people for their help with both the first and second editions of this book. My colleagues and students at California State University, Fullerton provided many provocative ideas. My friends and ex-colleagues who are still in the cable TV business helped bring me up to date on latest developments. My family members acted as a clipping service for me, providing me with articles whenever they came upon them.

I am also appreciative of the many helpful comments and suggestions provided by the reviewers: Thomas M. Irons, Washington State University; Roger Srigley, Michigan State University; Charles Steinfield, University of Houston.

The new television technologies continue to be exciting and intriguing. I hope this book will convey both a sense of enthusiasm and understanding to those who read it.

Lynne Schafer Gross

# part 1
## introduction

# overview

**1**

## Scope of the New Television Technologies

Just what are the new television technologies? This is not an easy question because television, itself, is a technology which fosters new developments almost on a daily basis. In recent times the term "new television technologies" has been used primarily to refer to new methods for bringing video signals to the home TV set. In this context, the new television technologies primarily have involved the distribution of televised material intended to be viewed by a sizeable audience. However, the television production area also has an illustrious history of technological innovation.

### Production Technologies

The early experiments with television began in the late 1800s and involved cameras and TV screens that were both crude and bulky. Picture resolution was poor, and the lights needed to create an acceptable image were enough to wilt the heartiest of souls.[1]

Gradually these problems were solved. By the late 1940s black and white studio cameras and TV screens were of acceptable quality, and TV rapidly became a widespread national phenomenon.

Through the years engineers developed new and exciting advances for the infant medium of television. Videotape recorders enabled programs to be taped and played back at a later time, greatly increasing the flexibility of the programming function while color added a new dimension to the picture.

Cameras light enough to be carried on the shoulder changed the whole news gathering function of television. News could be covered live on the spot without having to wait for film to be developed and without the logistical problems involved with large trucks and bulky cameras. A unit containing both a camera and a videotape recorder further enhanced the portable nature of TV. This was possible, in part, because videotape recorders had become smaller. The first large, bulky machines used tape that was two inches wide. Gradually the machines became more compact and tape sizes of one inch, ¾ inch, and ½ inch could be utilized and still give clear pictures.

Microprocessors and other electronic components increased the efficiency of the production process. The special effects possible through television switchers multiplied with each new model. The ability to play back material in slow motion, or even super slow motion, added interest, especially to sports programming.

Most of these technological breakthroughs helped the production process rather than the distribution process. They added interest, convenience, depth, or breadth to the methods of producing programming material, but they did not change how those programs reached the ultimate consumer.[2]

### Distribution Technologies

From the 1940s to the 1970s, the primary method for distributing television programs involved antennas and airwaves. Local stations sent signals from their antennas through the airwaves to rooftop antennas which sent the signals to home TV sets. If the signals needed to travel long distances, they were sent by microwave or long distance wires.

In the 1970s and 1980s, new methods of distribution came to the fore. Many of these were based on developments which were not technologically feasible before the 1970s—developments which involved satellites, wires, computers, and frequencies.

For example, the multitude of channels available on cable TV is possible because signals can be carried by satellites and then distributed by wire. Videotext material is possible because computers can hold large banks of information. Low-power TV is available because new equipment developments minimize the interference characteristics of broadcast frequencies.

This book will concentrate on these new distribution technologies—nontraditional methods by which video pictures arrive on the home TV set. Most of these are products of the late 1970s and early 1980s. Although some of them were developed almost as early as the traditional antenna-airwaves method, they did not receive widespread attention during their early years.


## Interrelationships

The new television technologies interrelate in many ways with each other, with established media, and with society. Throughout this book a number of recurring interrelationships will surface which have global, national, and local implications. The following are some of the major points to consider.

### International Implications

Radio waves do not obey boundary lines. A signal sent out from a town in Missouri might very well reach into Kansas, and a signal from Germany can be seen or heard in France. Disputes arise between the United States and its neighbors over signal strengths and permeation of signals. The Canadians who do not wish their country to be bombarded with American programming try to restrict border signals. The Mexicans frequently substantially increase signal strength on their own stations and, as a result, interfere with American stations that are on the same frequencies. In a similar manner, Cuba causes interference on U.S. stations.

International laws and international organizations such as the International Telecommunication Union, which sponsors regular World Administrative Radio Conferences, exist to try to settle such disputes, but often they are not successful because they have no real enforcement power.[3]

The new technologies have been particularly fraught with international implications. Some of the technologies took hold in foreign countries long before they were "discovered" in the United States. Britain and France have had teletext for years, and Canada was wired for cable TV long before the franchising boom hit the U.S. The political and economic relationships that arise between the United States and other world powers when technologies are transported in whole or in part from one country to another can be complicated and unwieldy.

Even more complicated is the process of trying to plan for compatibility and fairness in the new media. Superimposing worldwide technical standards on numerous already-existing standards is difficult, to say the least.

Space for television signals is not unlimited. Nations are constantly vying for spectrum space for shortwave radio and TV, broadcast radio and TV, and mobile radio. Because the number of frequencies available for these services are finite, each country wants to make sure it will have a large enough piece of the pie to accommodate present and future needs. An important international controversy involving the new technologies involves space for satellites. Only certain places in the sky are valuable for satellite placement, and the third world countries want to make sure these places are not all taken by the superpowers.[4]

Because of the international implications of the new technologies, many of the decisions involving American telecommunication services will not be made solely by the United States, but will be made at international conferences.

## New Technology Overlap

The various new television technologies compete and interact with each other on many levels. In some instances various delivery methods are trying to attract the same audience members. People do not have infinite time to spend with television, so they make choices regarding the media in rational and sometimes irrational manners.

The technology form which can approach people first with the shiniest image may capture the audience even though some other technology is cheaper and just as effective. Once a new form is established, it must always be looking over its shoulder because newer and shinier forms may be trying to lure away the audience.

Television technologies compete not only for audience but also for programming. Only a limited amount of truly worthwhile programming fare is created, and sometimes many of the different distribution forms desire it. This is particularly true of movies which are the mainstay of many programming services.

There are not enough good films made to fill the voracious appetite of the new media forms, and consumers tire of seeing the same material promoted over and over.

Many of the new television technologies are experiencing economic downturns due in part to overpromises that they made, and to the large number of them. A shakedown period will no doubt lead to survival of the fittest.

## Traditional Technologies

Each time a new entertainment mode takes hold, it alters former entertainment methods. Movies affected vaudeville; television affected radio; and now the television technologies are affecting the time-honored broadcasting structure.

For many years three networks, NBC, CBS, and ABC, dominated the programming function. Their affiliated stations were the most watched stations, and their programming was what was talked about each day in lunchrooms, school cafeterias, barber shops, and other gathering places. Even the independent stations survived primarily by rerunning old network shows.

The overall system ran one direction—from the networks and stations to the viewer. The time function was also controlled by the program providers. Viewers wishing to see "Gunsmoke" or other favorite programs knew they must be poised in front of their TV set at a particular time. The type of programming could be fairly well predicted—soft informational programs in the morning, game shows and soap operas in the afternoon, news around dinner time, situation comedies and drama in the evening, talk shows in the late evening, cartoons on Saturday mornings, religious programs on Sunday mornings, and sports heavy on weekend afternoons. Technological developments from the 1940s to the 1970s improved this structure but did not change it in any basic way.

In the 1970s technological advances which led to a myriad of program distribution methods began to progress geometrically, and new breakthroughs led to new services and new ways of looking at that old television set. Cable television with its multitude of channel possibilities began breaking down the old programming formats. Instead of a potpourri of programming that changed throughout the day, cable offered twenty-four hours of particular types of programs. Increased channel capacity led to the potential for increased choice on the part of the viewer.

Technology put the time function in the hands of the viewer through videocassette recorders, which could record programs off-air while the owner was out. Pre-recorded videocassettes and videodiscs allowed viewers to choose not only the time they wished to view but also, within limitations, the material they wanted to see. People could begin to view their TV sets differently than they had traditionally. TV viewing no longer needed to be a passive activity designed by programmers and absorbed by the audience. Individuals could now exercise some active control over the output of the glowing box situated in the living room.

The one-way nature of television continued to dissolve as interactive TV was introduced in cable systems and as videotext services became available. People began interfacing their TV sets with computers capable of retrieving information or simply playing games. This furthered the active rather than passive aspects of the TV set.

All of these structural changes led to a decrease in the influence of the three dominant networks—NBC, CBS, and ABC. The audience share which they enjoyed began to erode. Citizens who had greater choice of programming options offered exercised this choice by watching cable TV, subscription TV, or videotape cassettes. During the 1960s network primetime programs generally were required to attract twenty percent of the American TV households in order to be continued on a network. By the 1980s, the percentage needed had dropped to seventeen percent, primarily because pepole had not maintained their past loyalty to the networks.[5]

Yet the networks did survive, contrary to the opinions of some of the doomsayers of the 1970s. The glamour and proliferation of the new distribution methods lost their shine after a period of time. The old networks and the new distribution methods sought their own levels of success, a process still in transition.

### Regulation

As the new technologies evolve, attitudes toward regulation are also evolving. Federal regulation of traditional broadcasting began because of limited spectrum space. During the 1920s, when radio was unregulated, stations were interfering technically with each other putting the airwaves in total chaos. The government stepped in and established the Federal Communications Commission to prevent this interference.

Over the years, the FCC expanded its regulatory powers into other aspects of broadcasting such as ownership and programming. The justification for this has been scarcity of resources. The FCC has contended that the total number of radio and TV stations available is limited and, therefore, a government body should guarantee that existing stations are operated in the public interest, and for its convenience.

In its regulatory actions, the FCC has treated broadcasting as a local medium. Stations are licensed to local communities with requirements to serve local interests and needs. The networks, which are obviously predominant in programming fare, are not regulated. They receive regulatory pressures only through the local stations which they own or with which they are affiliated. The FCC holds the stations responsible for the programming which the networks distribute.

As the new television technologies are developed this regulation process is changing. The FCC continues to exert regulatory powers over radio and TV stations and over the entities of the new television technologies using radio waves. But, many of the new technologies do not send signals through the air. Cable TV uses primarily wire; video cassette recorders and video disc players are connected directly to home TV sets. For some of the new technologies no regulation exists at all. For others it has been relegated to a government group other than the FCC.

The scarcity of resources argument for regulation is rapidly disappearing. The new technologies provide many avenues for reaching people with televised messages. No longer is the variety of television available limited by the number of signals which can travel through the airwaves. This is causing the FCC and the television industry to reevaluate the amount of regulation needed to provide for the public interest.

The localism of regulation is also being reviewed. Local programming serves the needs of small groups of people, as opposed to national broadcast network programming which is aimed at the masses. But, with the diverse program base emerging through the various new media, small groups of people can be served with nationally based programming. Many of the programming services beamed to satellites for distribution by cable TV and other media are intended for groups with specialized interests. So, although the new technologies can provide a base for satisfying more individual tastes, television, itself, may be less local because much of the specialization comes through national or even international programming.

These changing views of regulation are interwoven with the political fabric of the time which emphasizes deregulation in all walks of life, not merely in the television sphere.

## Effects on the Poor

Unlike traditional broadcasting, most of the new media involve subscriber fees. People pay very directly for cable TV hook-ups, subscription TV services, video cassettes, videotext information, and many other services offered. These services are available only to those who pay, as opposed to broadcast services which are available to anyone with a TV or radio.

Under the broadcasting system the rich and poor had access to the same information and entertainment material. When fees are involved, the rich are able to subscribe to much more than the poor. Because many of the services in existence and proposed provide information, the rich will be wealthier in information than the poor. This could further the gap between the rich and poor in terms of knowledge, education, and social mobility.

## Privacy, Piracy, and Pornography

Several other issues are recurring themes among the new television technologies—the individual's right to privacy, the illegal duplication and selling of programming, and the showing of sexually explicit material.

The potential for invasion of privacy exists primarily because of the incorporation of computers in many of the new television technologies. If someone makes purchases or banking transactions through a TV set, they are recorded by a master computer. Likewise, data concerning programming that a person watches through pay-per-view or cable TV can be stored in a computer. An individual or group with access to this data could find out much more about a person than that person might wish revealed.[6]

A large illegal industry has sprung up around the new television technologies involving the bootlegging of programming material. Estimates are that the film and television industries are losing about a billion dollars a year to piracy.[7] Many different technologies exist for illegal distribution and reception of programs from the dubbing of videocassettes to the decoding of scrambled subscription TV signals. Although some of the piracy acts are clearly illegal, others are only questionably illegal or perhaps not illegal at all. Some supposedly illegal methods of distribution have been taken to court and received contradictory rulings as to their legality.[8] The whole issue is further clouded by what individual people do in their private homes. A person who videotapes a program off the air and then shows it to several friends is in a grey area of legality.

The traditional media, because they are mass media with a family orientation, have rarely been plagued with charges of pornography. Some of the new television technologies, however, have been much less restrictive in program content and, as a result, have come under fire from numerous citizens' groups. Some of these groups feel all sexually oriented material should be banned and others feel more precaution needs to be taken regarding its ease of access. The producers of the material reply that suppression of such programming is a violation of the first amendment. In addition, they argue, the sexually explicit material is shown on a restricted basis; people have to make a conscious purchase through such methods as pay-per-view, videocassettes, or subscription TV. The material does not come into their homes without their knowledge or desire.[9]

These issues and other sociological ones haunt the new television technologies. However, any time new developments appear, they are followed by new problems, some unforeseen. Generally, the societal structure deals with them in an open and fair manner and they are resolved or at least mitigated.

## Organization of the Book

This book is divided into four parts with Part 3 the longest because it deals with the specific forms of new television technologies.

### Part 1
Part 1 of the book is designed to give the reader basic background. This first chapter introduces the concept of new television technologies and lays the framework for the remainder of the book. The second chapter covers the traditional distribution processes which have been used most commonly to transport video signals from their point of origin to the home TV set. These processes, and newer ones, are utilized in a myriad of forms by the various new television technologies.

## Part 2

Part 2 gives special attention to two technologies which have played a very large role in making the new television technologies possible—satellites and computers. Although satellites can be considered a distribution method in their own right, they mainly serve as a necessary technological relay for many of the new television technologies which actually reach the consumer's TV set. Computers, which have their own fast-paced successful history, are an intricate part of most of the new television technologies.

## Part 3

Part 3 covers the new TV technologies, themselves. The first chapter in this section deals with cable TV—the most widespread and glamorous of the new forms. The next three chapters explain technologies which utilize the antenna-airwaves configuration but in a different manner than regular broadcast TV. Subscription TV scrambles signals sent from regular antennas; low-power TV uses regular broadcast frequencies but with much less signal strength; and multichannel multipoint distribution service uses frequencies that are higher than the regular broadcast frequencies. The following chapter covers satellite master antenna TV systems, set-ups that are akin to small cable TV systems located primarily in apartment complexes. The chapter on direct broadcast satellite discusses a method by which homes may receive signals directly from satellites without an intermediary such as a cable TV service. This is followed by two chapters on videocassette machines and videodiscs players, devices which within a consumer's home give flexibility in program viewing. The following two chapters cover teletext and videotext which are services that do not display full moving video pictures but rather display words, numbers, and graphics. They are primarily consumer information sources. The final chapter in Part 3 discusses technologies which primarily enhance reception on the TV set.

## Part 4

The one chapter in Part 4 gives some observations concerning the implications and interrelationships of the various new television technologies as they relate to each other and to other areas of society.

Overall, the book provides readers with an understanding of new television technologies and an ability to make wise decisions regarding their use, both in terms of personal applications and utilization by society as a whole.

# traditional distribution processes

**2**

## The Electromagnetic Spectrum

The basic distribution process used to transport video signals from their point of origin to the home TV set involves use of the electromagnetic spectrum. This spectrum has been used primarily to transport radio and TV station signals from the station antennas to the home rooftop antennas. The spectrum has also been used for microwaving signals across large distances. For many years, this was a primary method by which networks distributed programs to stations.

Another traditional method of distribution involves the use of wires. This has been used primarily for long distance transportation of signals. Distribution of programming can also be accomplished by more mundane means such as shipping programs through the mail or sending them by airplane, train, car or bus.

This chapter explains these traditional distribution methods and sets the stage for the discussion of their expansion and use in relationship to the new television technologies.

### Frequencies

The electronic spectrum is a continuing range of energies at different frequencies which encompasses much more than broadcast program distribution. These frequencies can be compared to the different frequencies involved with sound. The human ear is capable of hearing sounds between about 16 cycles per second for low bass noises and 16,000 cycles per second for high treble noises. This means that a very low bass noise makes a vibration that goes up and down at a rate of 16 times per second. These rates are usually measured in "hertz" in honor of the early radio pioneer Heinrich Hertz. One hertz (Hz) is one cycle per second. So a low bass note would be at the frequency or rate of 16 Hz, a higher note would be at 100 Hz and a very high note would be at 16,000 Hz.[1]

As the number becomes larger, prefixes are added to "hertz" so that the zeros do not become unmanageable. One thousand hertz is referred to as one kilohertz (KHz), one million hertz is one megahertz (MHz), and a trillion hertz is one gigahertz (GHz). So the 16,000 hertz high note could also be referred to as having a frequency of 16 KHz.

Radio waves which are part of what is referred to as the electromagnetic spectrum also have frequency and are measured in hertz. But their frequencies are

**Figure 2.1.** The electromagnetic spectrum.

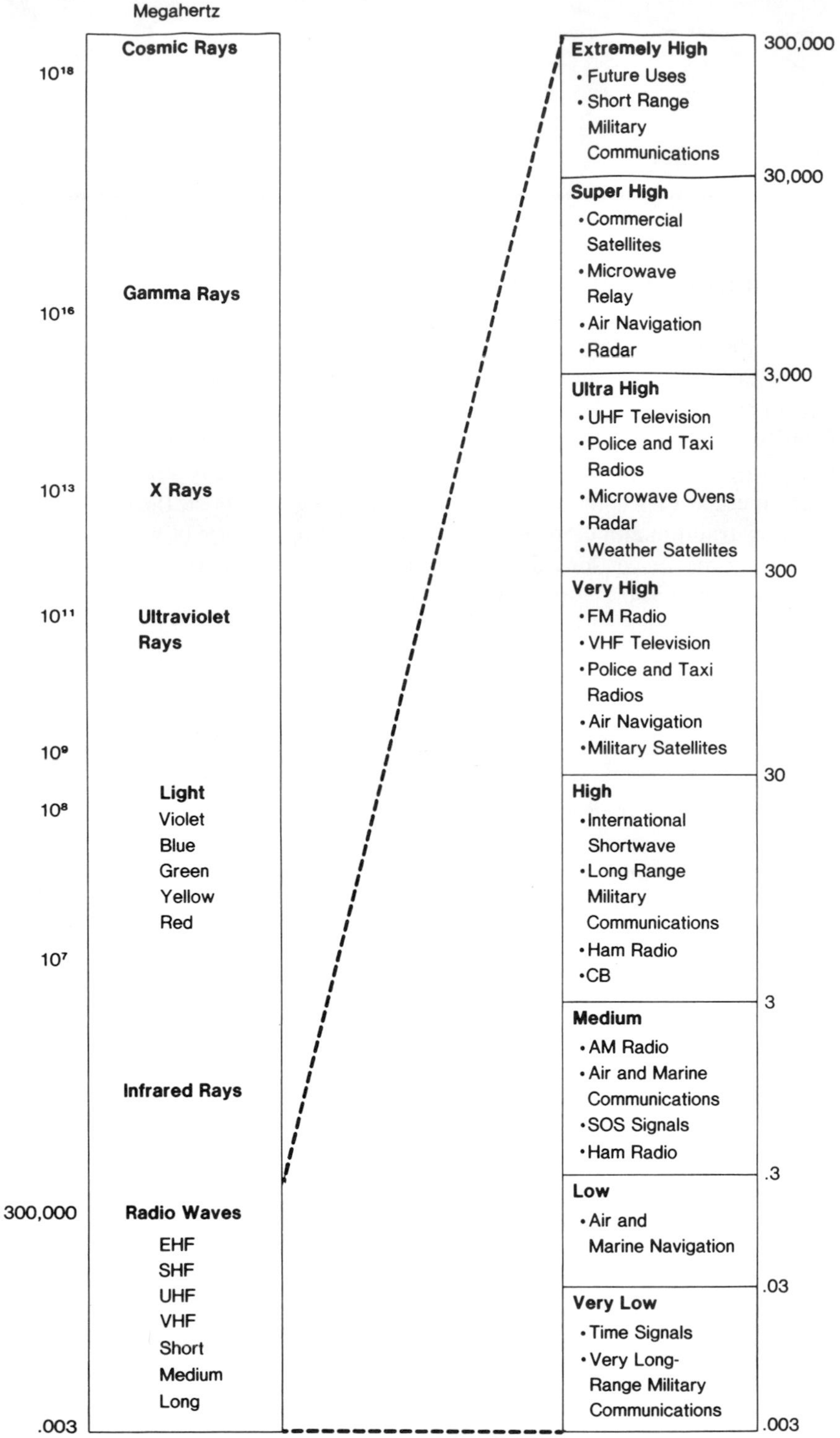

Megahertz

| | |
|---|---|
| **Cosmic Rays** | |
| $10^{18}$ | |
| | |
| **Gamma Rays** | |
| $10^{16}$ | |
| | |
| **X Rays** | |
| $10^{13}$ | |
| | |
| **Ultraviolet Rays** | |
| $10^{11}$ | |
| | |
| $10^9$ | |
| **Light** | |
| $10^8$ | Violet |
| | Blue |
| | Green |
| | Yellow |
| | Red |
| $10^7$ | |
| | |
| **Infrared Rays** | |
| | |
| 300,000 | **Radio Waves** |
| | EHF |
| | SHF |
| | UHF |
| | VHF |
| | Short |
| | Medium |
| | Long |
| .003 | |

| | |
|---|---|
| **Extremely High** | 300,000 |
| • Future Uses | |
| • Short Range Military Communications | |
| **Super High** | 30,000 |
| • Commercial Satellites | |
| • Microwave Relay | |
| • Air Navigation | |
| • Radar | |
| **Ultra High** | 3,000 |
| • UHF Television | |
| • Police and Taxi Radios | |
| • Microwave Ovens | |
| • Radar | |
| • Weather Satellites | |
| **Very High** | 300 |
| • FM Radio | |
| • VHF Television | |
| • Police and Taxi Radios | |
| • Air Navigation | |
| • Military Satellites | |
| **High** | 30 |
| • International Shortwave | |
| • Long Range Military Communications | |
| • Ham Radio | |
| • CB | |
| **Medium** | 3 |
| • AM Radio | |
| • Air and Marine Communications | |
| • SOS Signals | |
| • Ham Radio | |
| **Low** | .3 |
| • Air and Marine Navigation | |
| **Very Low** | .03 |
| • Time Signals | |
| • Very Long-Range Military Communications | |
| | .003 |

higher than sound waves and they can be neither seen nor heard but are capable of carrying sound. It is with these radio waves that most of the radio and television distribution takes place.

Above radio waves on the electromagnetic spectrum are infrared rays, and then light waves, with each color occupying a different frequency range. After visible light come ultraviolet rays, X rays, Gamma rays, and cosmic rays.

## Radio Waves

The radio wave portion of the spectrum is divided into eight sections, each comprised of a band of frequencies measured in Hz. Because most of the radio and television applications are in the megahertz range, these frequencies can best be discussed in terms of MHz.

.003 to .03 MHz are called very low frequencies.
.03 to .3 MHz are called low frequencies.
.3 to 3 MHz are called medium frequencies.
3 to 30 MHz are called high frequencies.
30 to 300 MHz are called very high frequencies or VHF.
300 to 3000 MHz are called ultra high frequencies or UHF.
3000 to 30,000 MHz are called super high frequencies.
30,000 to 300,000 MHz are called extremely high frequencies.[2]

Although all radio frequencies in the electromagnetic spectrum are capable of carrying sound, they are not all the same. The frequencies which are toward the lower end of the spectrum behave more like sound than the frequencies which are at the higher end. These higher frequencies, in turn, behave more like light. For example, the lower frequencies can go around corners better than the higher frequencies in the same way that you can hear people talking around a corner but cannot see them.

Radio waves also have many uses other than the distribution of entertainment and information programming usually associated with the television industry. These uses range from the opening of garage doors to highly secret reconnaisance functions.

## Service Placement

The spot on the radio portion of the spectrum where a particular service is placed depends somewhat on the needs of the service. The lower frequencies have longer ranges in the earth's atmosphere, so long range military communications appear on the very low frequencies while short range appears on the extremely high frequencies.

Many placements, however, are an accident of history. The lower frequencies were understood and developed earlier than the higher frequencies. In fact, in early days people were not even aware that the higher frequencies existed, and today the extremely high frequencies are still not used to any great extent. AM radio was developed earlier than FM so was placed on the part of the frequency that people then knew and understood. The "discovery" of ultra high frequencies during World War II led the FCC to reallocate television frequencies after the war into two categories, VHF and UHF.

The portions of the electromagnetic spectrum which are of greatest importance to the distribution of entertainment and information programming connected with the traditional broadcasting stations are the following:

.535 MHz to 1.605 MHz—107 AM radio channels (This band of frequencies is usually referred to as 535 to 1605 KHz)
54 MHz to 72 MHz—TV channels 2, 3, and 4
76 MHz to 88 MHz—TV channels 5 and 6
88 MHz to 108 MHz—100 FM radio channels
174 MHz to 216 MHz—TV channels 7 to 13
470 MHz to 890 MHz—TV channels 14 to 83

Some of the distribution methods which come under the label of new television technologies use the frequencies listed above. For example, subscription TV uses regular UHF channels 14 to 83 but scrambles the signal. Low-power TV stations occupy the same frequencies as regular VHF and UHF stations but broadcast with less power.

Other new television technologies use frequencies that are much higher than the radio and TV station range, in part because they were developed after all the lower frequencies had been taken. For example, multichannel multipoint distribution service uses frequencies in the 2500 MHz range and Direct Broadcast Satellite is above 12 GHz.

The electromagnetic spectrum is important to the whole of television—both the traditional and new technology forms.[3]

## Radio Station Broadcasting

The first of the mass media to distribute programming using the electromagnetic spectrum was radio. Therefore, many of the methods of spectrum use were developed through radio and then transferred to broadcast TV and then to the new television technologies.

Radio has two different transmission modes, AM and FM. AM broadcasting is located in the medium frequencies and FM in the very high frequencies. The position in the spectrum, however, has nothing to do with AM (amplitude modulation) and FM (frequency modulation).

### Modulation
Modulation refers to the method by which sound is placed on a radio frequency carrier wave. The sound created by a radio station reaches that station's transmitter and antenna in the form of variations in electrical energy. This electrical energy is then modulated, which means it is superimposed onto the carrier wave

which represents that particular radio station's frequency. The sound energy cannot go through the air itself because it does not move fast enough. It must be carried on a radio wave which is of much higher frequency.

The transmitter generates this carrier wave and places the sound wave on it by a process called modulation. This modulation can occur as either amplitude modulation or frequency modulation regardless of where the carrier wave is located on the spectrum. In amplitude modulation, the amplitude, or height, of the carrier wave is varied to fit the characteristics of the sound wave. In frequency modulation, the frequency of the carrier wave is changed instead.[4]

## AM and FM Characteristics

AM and FM have different characteristics caused by the differing modulation methods. For example, FM is static-free while AM is subject to static. This is because static appears at the top and bottom of the wave cycle. Because FM is dependent on varying the frequency of the wave, the top and bottom can be eliminated without distorting the signal. AM, however, is dependent upon height, so these static regions must remain with the wave.

AM and FM also have differences due to their placement on the spectrum. AM signals can travel great distances around the earth while FM signals are nearly line-of-sight and often cannot be heard if a building or hill comes between the transmitter and the radio attempting to receive the station. This is because FM is higher up on the spectrum, closer to light than AM. Light waves do not travel through buildings or hills and FM signals are similarly obliterated.

AM, at the lower end of the spectrum, is not so affected, but these lower frequencies are affected by a nighttime condition of the ionosphere which, when hit by radio waves, bounces the waves back to earth. In this way AM waves can be bounced great distances around the earth's surface—e.g. New York to London. FM is not affected by this layer because of its position in the spectrum, not because of its manner of modulation. Theoretically, if FM waves were transmitted on an AM frequency such as 535 kilohertz, they could bounce in a manner similar to AM.

## Bandwidth

Another difference between AM and FM is the greater bandwidth for FM stations. Although each station is given a specific freqency such as 550 kilohertz or 88.5 megahertz, the amount of spectrum actually used covers a wider area. For an AM radio station this width (bandwidth) is ten kilohertz and for FM the bandwidth per station is 200 kilohertz. So an AM station at 550 kilohertz actually operates from 545 to 555 giving it room to modulate the necessary information and prevent interference from adjacent channels. Because FM stations have a broader bandwidth, they produce higher fidelity and contain more information than an AM station.

## Stereo

Broader bandwidth makes FM more adaptable to stereo broadcasting. An FM wave has enough bandwidth to carry more than one program at a time. Therefore, part of the wave can carry sound as it would be heard in the left hand side of a concert hall and another part of the wave can carry sound as it would be heard in the right hand side. The result is a stereo transmission.

A method has been engineered for transmitting AM stereo. In fact, several methods have been engineered but they are not compatible with each other. In April of 1980, the FCC selected one of these, a system designed by Magnavox, to be the industry standard. The other four companies which had developed AM stereo at the time—Harris, Belar, Kahn, and Motorola—objected, along with several unbiased industry practitioners who also questioned the Magnavox choice. As a result, the FCC failed to reconfirm its decision regarding Magnavox, and instead decided to let the market place resolve the situation.

This left individual stations in a quandry. If a station chose one method and bought all the equipment necessary to broadcast AM stereo in that method, it might be left out in the cold if another method became the favorite of most radio stations and the consumers who purchased sets to receive another brand of AM stereo. Sets compatible with all forms of AM stereo can be made, but they are very expensive, so multiple AM stereo will probably not be the answer. The field of companies still interested in AM stereo has narrowed somewhat with Kahn and Motorola being the only two still actively seeking acceptance. Of these two, Motorola has the most stations signed up to use its system.[5]

## Multiplexing

The broader bandwidth of FM enables independent signals other than those needed for stereo sound to be multiplexed on an FM radio station signal. This is referred to as SCA or sub-carrier authorization. The information is carried on the FM signal but can be received only by special receivers. Under the original SCA rules, commercial FM stations had to offer broadcast-like services such as music piped into doctors' offices. Noncommercial FMs could offer only educational services, such as reading for the blind, on a non-profit basis. In 1983 a rule changed allowing both commercial and noncommercial FMs to carry all kinds of communication services on a for-profit basis. Some of them have entered into fields related to the new television technologies such as electronic mail, paging, and dispatch.[6]

## Reception

After the sound waves have been modulated either AM or FM onto the proper frequency carrier wave at the transmitter, they are radiated into the air through the radio station antenna at the assigned frequency and power.

These waves are always in the air but they do not distribute program material unless someone turns on a radio receiver. The purpose of the receiver is to pick up and amplify radio waves, separate the information from the carrier wave, and reproduce this information in the form of sound. When radio was first invented, a huge console, which usually sat in the living room, was used to accomplish these steps. Now, thanks to technological advances, all this can be accomplished in the smallest of receivers worn in the ear, on the wrist, or in a pocket.

Although broadcast television and the new television technologies undergo transmission and reception processes that are more complicated because they involve both video and sound, many of the broad processes are the same as those which have been developed for radio.

## Television Station Broadcasting

With television as with radio, it is at the transmitter that information is super-imposed or modulated onto a carrier wave. For television, audio and video are sent to the transmitter separately—the video signal is amplitude modulated and the audio signal is frequency modulated. The two are then joined and broadcast from the antenna.

### Stereo
At present, most TV audio is monaural, but a method for broadcasting TV sound in stereo has been developed. It is called multichannel television sound or MTS. Some stations are considering using the second channel for a foreign language soundtrack rather than stereo. Before MTS can become widespread, stations must invest in the equipment to produce and broadcast two channels and consumers must purchase sets which have been manufactured to receive two channels.[7]

### Bandwidth
Again, as with radio, each TV station has a frequency band on which it broad-casts. A television station, however, uses a great deal more bandwidth than a radio station. While AM stations span 10 KHz and FM stations 200 KHz, the bandwidth for a TV station is 6000 KHz or 6 MHz. This means that a TV station takes 600 times as much room as an AM station and that all the AM and FM stations together occupy less spectrum space than four TV stations. The reason for this is that video information is much more complex than audio information. More room is needed to modulate video information and to protect the modulated information from interference.

### Spectrum Placement
TV channels are placed at various points in the spectrum. There is a small break between VHF channel 4 and channel 5 and a larger break between channel 6 and channel 7 that encompasses, among other services, FM radio. In general, in any particular area two adjacent channels cannot be used because they would interfere with each other. For example, Detroit could not have operating stations on both channel 2 and channel 3. However, because of the break in spectrum

between 4 and 5 and between 6 and 7, communities can use those adjacent channels. For example, New York has a station on both channel 4 and 5.

Channels 14 to 83 are well above channels 2 through 13 in the UHF range. Because these channels have not been highly utilized, the FCC does not generally allocate channels 71 to 83 for broadcast purposes.

Both UHF and VHF waves follow a direct, line-of-sight path, but UHF signals, because they are closer to light, are more easily cut off by buildings and hills and are more rapidly absorbed by the atmosphere. Because of this they require higher power at the transmitter to make up for these losses. Both VHF and UHF antennas are most effective if located at a high point.

**Multiplexing**
Multiplexing on a TV station signal is not as easy as multiplexing on an FM station signal because of the greater complications and requirements of TV. This process is one that comes under the new television technologies in the form of teletext. Words and numbers are broadcast on part of the TV station frequency and can be received with special decoders.

**Reception**
Once a TV signal is sent from an antenna, it must be received in order for the distribution process to take effect. The signal is picked up by an antenna, amplified, and then reproduced as picture and sound. The audio is produced in a manner similar to that of radio.

Video is produced by electronic devices called guns located in the rear of the cathode-ray, or picture tube. A black and white set holds a single gun. A color set contains three guns, one for reds, one for blues, and one for greens. The incoming video signal causes the gun (or guns) to send out electrons which make the phosphor-coated screen at the front of a TV set glow. A black and white set glows in blacks, whites, and grays according to the intensity of the signal. A color set glows in various combinations of red, green, and blue. The television picture is actually an array of small glowing dots blinking very rapidly in various degrees of brilliance. The beam from the electron gun (or guns) starts in the upper left hand corner and activates the phosphor one dot at a time from left to right. It does this all the way down the screen for 525 lines. A TV picture is really a series of dots which the eye perceives as a moving picture.[8]

The principles behind the broadcast television transmission and reception process have been well developed over the years. Many of these principles have been utilized in the development of the new television technologies.

## Wire Transmission

Broadcasting utilizing a transmitter and antenna is fine for disseminating programming in a local area. If the programming is to go a great distance, however, some other form of distribution must be used.

## Phone Wires

One form of distribution is wires. Special phone wires giving higher fidelity than regular telephone wires can be used for audio transmission. Radio network programming can be sent across the country on these wires. Videotext information, wire service news headlines, and computer data bank information, all consisting of letters and numbers can also be sent on wires to cable systems or other places wishing to receive the information. The wire system is augmented by amplifiers that keep the signal strong. AT&T owns most of the phone wires across the country, and networks or other entities which wish to use the wires lease time on them from AT&T.

## Coaxial Cable

Phone wires are not adequate for television signals because they can not carry the amount of information needed for video. A special type of wire called coaxial cable has been developed for television and is used extensively in many phases of TV. It connects cameras to other production equipment, connects videocassette recorders to TV sets, and interconnects entire cable TV systems.

## Fiber Optics

Looming on the horizon is the use of fiber optics which could completely revolutionize wire transmission. Fiber optics involves using very thin strands of glass as a carrying device for audio and video signals. The strands are capable of carrying much more information in much less space than present copper wires. Each glass fiber is less than one hundredth of an inch thick or about the thickness of a hair. In general, one pound of fiber optics can carry eighty times more information than one pound of coaxial cable.

Within the strands of glass, the signals are carried on laser light produced by laser diodes which are very closely related to the light-emitting diodes (LEDs) in digital watches. Laser diodes emit infrared light which passes through the glass fibers. This is advantageous because infrared light travels further through glass than other forms of light. Generally, light travels only in a straight line, but light contained in a fiber can be reflected around corners by the glass walls.

Another advantage of fiber optics is that light sent through glass is less likely to encounter interference than light sent through copper wire because glass is not subject to electromagnetic energy. Fiber optics is being used, primarily in situations that require new wiring. Its takeover of traditional wiring systems will be slow but steady.[9]

Wire transmission is generally reliable, but it requires forethought because the wires need to be strung from one location to another. This is in contrast to distribution methods using the electromagnetic spectrum which flow freely connecting distant areas.

# Microwave

Another disadvantage of wires is that they are not always the most effective means to accomplish long distance distribution of programming material. Installing cable in mountainous, uninhabited, or irregular terrain is not efficient either in terms of initial installation or maintenance.

### Relay Links

For many years the common method for distributing television network signals across the country was through a microwave relay link. Microwaves are very short waves higher in the spectrum than radio or TV station allocations. Their line-of-sight quality demands that relay stations be in sight of each other and not more than about thirty miles apart. Each microwave station across the nation is mounted on a tower or tall building in a relatively high place. The first station in the chain sends the network signal to the next station which receives the signal, amplifies it, and sends it on to the next station. The last station need only receive the signal.[10]

AT&T owns the microwave links, so the networks and other users of the microwave links lease time from AT&T. For this reason no network relays material indefinitely on any particular microwave frequency. AT&T decides which frequencies are available for the various customers. In general, microwave relays are common in rural areas while coaxial cable distribution is common in urban areas.

### Uses

Microwave can be used for purposes other than network feeds. Stations often place a microwave dish on the top of a remote truck to send the signal back to the station. Cable TV companies sometimes connect different parts of their systems through microwave interconnects.

Microwave use is still quite common for both traditional and new television technologies, but some of its duties are now performed by satellites.

# Physical Transportation

Many forms of program distribution involve the physical transportation of a program from one place to another. Usually this means that one or more human beings on foot or in some sort of vehicle actually pick up a tape and carry it. One term used to describe this type of distribution is bicycling, because at one time messengers on bicycles carried material from one point to another.

### Old Uses

This form of distribution was frequently used in traditional television. For many years program syndicators who sold material to independent stations mailed or carried programs from one station to another. Likewise, educational television at one time had an elaborate bicycle network.

## New Uses

Although the new technologies boast many ethereal methods of distribution, they still employ physical transportation. Video discs and videocassettes are delivered to video stores by trucks and are then picked up by consumers who come to the stores on foot, in a car, or perhaps even by bicycle.

Although the pick up and carry method of distribution is not as glamorous as microwave, coaxial cable, or broadcast transmission, it is often the most effective and least expensive means of distribution.

# part 2
## foundations

# satellites

# 3

## Description

Satellite technology is one of the most important underpinnings for the new television technologies. Although the traditional distribution methods such as microwave, wires, and broadcasting are important to many of the new technologies, it was the advent of satellites that made possible their widescale development. Satellites provide an easy, flexible, relatively inexpensive method for transmitting information from one point to another.

### Positioning

Communications satellites are launched using rockets, in the same manner as space exploration satellites are launched. Launches are designed so that satellites are positioned 22,300 miles above the equator where they can travel in an orbit that is synchronized with the speed of the rotation of the earth. In this way these synchronous satellites appear to hang motionless in space and can continually receive and send signals to the same points on the earth. When the satellite is in orbit, is is powered by solar energy it gathers from the sun.

The satellites that transmit to the United States are positioned along the equator between 55° and 140° longitude. From there they can cast a signal, called a footprint, over the entire United States.[1] Other satellites positioned at other longitudes have footprints over other sections of the world. Most of the surface of the globe can be covered by three satellites strategically placed.

Instant worldwide communication is possible through a worldwide satellite network, and pictures can be beamed from anywhere in the world to the United States. In parts of the world where countries are small and close together, signals intended for one country can easily be picked up by other countries.

### Transmission

Signals are sent to satellites by large ground station dishes, which have transmitters that electronically place the information to be sent on radio waves in the gigahertz range, much higher frequencies than those used by conventional broadcasters. Two main frequency bands have been set aside for satellites, four to six gigahertz, which is referred to as the C-band, and eleven to fourteen gigahertz, which is referred to as the Ku-band.[2] The ground station dishes (also referred to as earth stations or as an uplink) are positioned so that the signal they send will be received by the appropriate satellite.

The waves travel through the earth's atmosphere where they experience some interference, but once into outer space they do not encounter inhospitable conditions or bad weather so are virtually home-free for their trip to the satellite. In this way satellite links are advantageous over earthbound forms of television communication distribution that have been used in the past, primarily microwave and coaxial cable.

## Transponders

Once the signal arrives at the satellite, it is placed on a particular transponder. Each satellite contains about twenty-four transponders, which are roughly analagous to channels. A satellite with twenty-four transponders, theoretically, could receive twenty-four different program signals sent from twenty-four different earth stations, all pointed at the satellite. In reality some of the transponders are used for internal communication to keep the satellite on course and to make sure all information is properly placed. Most communication satellites carry eighteen to twenty-two different programming sources.[3]

## Reception

Satellites transmit their signals back to earth, first through outer space, and then through the earth's atmosphere. These signals are picked up by ground station dishes, which look similar to the sending dishes. However, receive only dishes are much less expensive than sending dishes because they are not attached to transmitters. Again, these ground stations are often referred to as earth stations or as downlinks.

Anyone with a ground station dish properly positioned to pick up satellite signals can receive the programming on all the transponders. This, once again, is an advantage satellite transmission has over coaxial cable or microwave. Both of those distribution systems take signals from one point to another single point, e.g. from the beginning of the wire to the end of the wire or from one microwave dish to another. Having more than one reception point adds to the cost and often degrades the signal. But with satellite transmission, an infinite number of dishes can pick up the signal without any degradation of quality to any of the reception points.

Most satellite receiving dishes can pick up either all the satellites in the C-band or all the satellites in the Ku-band. The dishes needed for C-band measure about ten feet across while those needed for the Ku-band are only about three feet.[4]

Cable TV systems are the main organizations receiving satellite delivered programming at present, most of which is transmitted in the C-band. Some individuals have bought earth stations which, when placed in yards or on roofs can receive programming intended for cable TV or other services. Many hotels and apartment complexes have also purchased their own satellite reception dishes.

**Figure 3.1.** A satellite and its footprint.
Photo courtesy of Hughes Aircraft Company.

Other new technologies such as MMDS and low-power TV stations are also designed to receive satellite signals. The more traditional broadcast stations and networks employ satellite transmission rather than microwave to send feeds around the country. Direct broadcast satellite is planned as a system to be received by individuals.

### Planned Improvements
Satellite technology will continue to improve. At present satellites are launched to positions four longitude degrees apart, but, as of 1987, some satellites will be launched to orbit two degrees, or approximately 900 miles, apart. This method of positioning will open up room for many more satellites.

The price of satellite dishes continues to fall. In 1980, a dish set-up cost approximately $10,000. Now, the price is below $1000 and continues to fall.[5]

As the space shuttle becomes more sophisticated and useful, it will be able both to build and repair satellites in space.

Communication satellites are definitely one form of the new television technologies which is here to stay. As technological improvements are made, new uses will be found for these birds in the sky.

## History

Satellites, themselves, are not new, but new uses for them are developed on a regular basis.

### Early Communication Satellites

The first communication satellite to be launched was Telstar I, sent up in July of 1962 by AT&T and the National Aeronautics and Space Administration. This satellite carried, among other things, the first live television transmissions between Europe and the United States. Before this there was no method for live television transmission from one part of the world to another. Laying coaxial cable across oceans was not practical and microwave could not span oceans, but satellite transmissions could span this distance.

However, Telstar I had a flaw in that it did not permit continuous transmission because it was not a synchronous satellite, and its signal could not be received when it dropped below the horizon on its elliptical orbit. The first synchronous satellite, Early Bird, was launched in April of 1965. Its orbit was timed so that it appeared to "hang in space" as all present day communication satellites do.

### COMSAT and INTELSAT

In order to control satellite communication, Congress passed a 1962 law establishing COMSAT, the Communications Satellite Corporation, which was set up as a private company with half the stock owned by the general public and half owned by communications companies. This company oversaw the early satellite development in this country.

COMSAT, with the help of NASA, launched several satellites named Comstar, and then other companies, most notably Western Union and RCA, asked for and received permission to operate satellites, which they called Westars and Satcoms, respectively. Later other companies such as Hughes, entered the satellite launch business.[6]

All of these companies received permission to operate from the FCC, paid NASA to launch their satellites, and then rented the facilities to smaller companies that transmitted to and from the various transponders. In this way these companies acted as common carriers in that they rented to anyone willing to pay for the service.

To determine international policies regarding satellites, an International Satellite Consortium (INTELSAT) was formed by most of the countries friendly to the United States. For over twenty years INTELSAT, owned jointly by over 100 countries, was the world's exclusive provider of international satellite links. In the 1980's, its monopoly was questioned by several American companies wishing to compete with INTELSAT. The Reagan administration, in a deregulatory mood, favored the development of competition for INTELSAT to assure that rates would remain low and service would be of high quality for international satellite functions. Not all of the other countries agreed and INTELSAT, itself, was not overly receptive to the idea. Several companies indicated interest in providing international satellite service and awaited a decision.[7]

## Early Uses

Satellites are, of course, used for many purposes other than communications including military applications and weather gathering. The early communication uses of satellites include transmitting of noteworthy international events such as Pope Paul's visit to the United States, splashdowns of U.S. space missions, track meets in Russia, Olympic games, and international talk shows. The broadcast networks used this material either as special programs or as inserts for news broadcasts.

Gradually, other uses surfaced for satellite interconnects. Networks began using them to send remote programming, such as a football game in Missouri, back to the network headquarters in New York where it could then be sent to affiliated stations through the network's microwave line leased from AT&T. In this same manner the Johnny Carson show was sent by satellite from Burbank where it was produced back to New York where it was sent by microwave back to KNBC in Burbank and, of course, to NBC affiliates in other cities.[8]

The Public Broadcasting System decided to use satellites to distribute all its programs to the public television stations, and National Public Radio also began transmitting its audio service by satellite.[9]

The Robert Wold Company leased large blocks of satellite time, then, acting as a broker, sold the time to companies who wanted to use satellites on a one-time or limited basis. For example, U.S. Tobacco, which specializes in the sale of snuff and chewing tobacco, identified a number of cities that it felt were prime potential sales areas. Through Wold, which rented it not only satellite time but also sending and receiving dishes, it joined together stations in those cities for the sponsored telecast of the Fort Worth National Rodeo Championship.[10]

## Teleconferencing

The Wold Company also pioneered in another use of satellites, which was to have increasingly widespread acceptance. This involved the concept of teleconferencing, also known as videoconferencing.

Private companies, corporations, or organizations would rent satellite time, dishes, and television equipment from Wold or other companies in the teleconferencing business in order to conduct meetings. Rather than flying employees across the country or around the world to discuss important business concepts, the corporations would set up a satellite interconnect and hold a meeting.

For example, a group of executives sitting in a hotel in Los Angeles would be covered by microphones and TV cameras that were in turn connected by wire to an earth station that transmitted to a satellite transponder. What the Los Angeles executives said and showed could then be sent from the satellite transponder to an earth station in New York. A wire from this earth station to a TV set in a hotel room in New York enabled New York executives to see and hear what the Los Angeles executives were discussing. If the New York hotel rooms were equipped with cameras and microphones, and if the New York earth station housed a transmitter as well as a receiver, then the New York executives could talk back to the Los Angeles executives and a two-way meeting could take place, which would be very similar to a meeting conducted if all the executives were in one room.

Companies found that such arrangements could save money because the tele-conferencing costs were less than the transportation, lodging, and meal costs of transporting executives around the country, especially when those costs also included lost time on the part of those executives as they wended their way to airports and sat on airplanes.

Similarly, trade organizations found that they could hold regional meetings, which, through satellite, could become national or international meetings. For example, the American Soybean Association held a conference that linked together Tokyo, Rio de Janeiro, London, and Atlanta through a satellite teleconferencing set-up. In this way approximately 1500 soybean growers were able to hear experts from four continents discuss market outlooks for their crops. Uplinks at the four cities enabled the remarks being made at each location to be fed to satellites that transmitted the messages to meeting halls at the other locations.[11]

## Cable TV Use
The early network and teleconferencing uses of satellites created some demand for satellite transponder time, but it was Home Box Office's placement of its pay-TV service on RCA's Satcom I in the mid 1970s that opened the floodgates. HBO was an infant company, which was essentially hand carrying movies from one cable system to another so that the cable systems could show the movies on an empty channel, collect money from the viewers, and in this way gain a little added revenue. HBO decided that distributing these movies by satellite would be easier and might increase its sales because any cable system that put up a satellite dish could receive the movies and then send them to subscribers over the cable. This concept did not meet with immediate success because cable systems were not willing to invest in satellite receiving dishes. However, after some aggressive marketing and technical and financial support on HBO's part, the concept took hold in a big way.

Many other companies followed HBO's lead and established programming aimed at cable systems, which they distributed over satellite. At first most of this programming only filled a few hours a day, but gradually twenty-four hour programming became the norm. With the cable industry taking to satellite, the demand for transponders well outstripped the supply.[12] When Satcom III, launched by RCA in December 1979, was lost, the cable industry companies that were planning to place their services on that satellite complained so bitterly that RCA leased time on other companies' satellites and re-leased it to the cable companies.[13] Gradually the number of satellites launched caught up with the demand for satellite transponders and the cable industry and other users were able to be accommodated.

### Backyard Reception
From the very earliest satellite transmission time, it was possible for anyone with a receiving dish to obtain the signals being transported by satellite transponders. When satellites were first developed, however, these dishes cost in the neighborhood of $50,000 dollars, so individual citizens were not likely to purchase them. Over the years the cost of dishes declined dramatically and came into the price range of some consumers.

At first the people purchasing dishes were the rich and the electronic tinkerers who were willing to pay a premium price to experiment with a new technology. But, as the price of dishes declined and the number of programming services on the satellites increased, more and more people began buying satellite dishes for their backyards or roofs. At first most of these purchases were in rural areas where TV reception was limited, but gradually the dishes entered more highly populated areas where other forms of distribution, such as cable TV, were available.

The people purchasing dishes could receive all the cable TV network signals as well as many of the transmissions which ABC, CBS, and NBC were sending across country. All of these were free for the taking once the dish was purchased.[14]

As the number of backyard dishes approached the one million mark, the cable services, in particular, became defensive. These dish owners were receiving HBO, the Disney Channel, ESPN, and other services for free, rather than purchasing them from the local cable company.

Several of the cable services, led by HBO, made plans to scramble their signals so that they could no longer be received in an intelligible form by the dish owners. These companies were planning to make the services available to the dish owners, provided they paid a fee to cover the descrambler and the programming.[15]

### Additional Uses
Other groups besides cable TV systems and backyard owners desired the programming transmitted by satellites. Apartment and condominium complexes that either did not want to wait to be cabled or did not want to deal with cable systems put up their own satellite dishes and began receiving the programming which they then sent by wire to the various units in the complex wishing to subscribe.

Hotels and bars also brought in satellite signals for their customers. In most instances, these types of users paid the cable programmers for the material.

Additional uses for satellite transmission emerged in the conventional broadcasting field simultaneously with the increased uses by cable TV and others. Syndicators of such programs as "The Merv Griffin Show" and "Hour Magazine" who previously had mailed or hand carried tapes to stations began simultaneously distributing by satellite and talking the stations who carried the program into buying satellite receiving dishes.[16]

Several new radio networks sprung up delivering their programming exclusively by satellite.[17] Established networks, both radio and TV, began converting from phone lines or microwave relays to satellite.[18]

Satellites, in their short history, have become an important method of program distribution. They have been placed in synchronous orbits with specific footprints, and their transponders can both receive and transmit to earth stations. Uses for satellites have gone from occasional special pick-ups to twenty-four hour program services. They may, in time, replace most long distance wire and microwave distribution. From all indications, the sky is the limit.

## Issues

The existence of satellites has created issues involving many of the other new television technologies, particularly as they vie in the competitive world for the consumer's dollar. For example, cable TV systems would like satellite signals restricted to themselves so other new media would not pose competition.

Satellites are by now accepted as a fact accompli, but there are major issues that need to be faced as the technology develops further.

### International Power

One of the most controversial of all issues surrounding satellites is the international allocation of satellite positions. There are only a limited number of prime orbital spots in the sky for synchronous communication satellites. If satellites are not placed by the equator at 22,300 miles, they cannot beam continuously to one area of the earth.

The United States, because it has the technology, is quickly filling the prime positions. Third world nations are uneasy about this and are pushing for reservation of satellite space for themselves until they have the capability of launching their own satellites. The U.S. argues that letting orbital positions lay fallow will slow down satellite technology for everyone, and that the places should be filled on a first-come, first-serve basis.

Programming on the satellites presents a particularly thorny international issue. Satellites cover a much broader area than conventional TV stations, and if individuals can pick up satellite signals in their own homes, they will have easy access to information that governments may not want them to hear.

Most of Europe, for example, can be covered by the signal of a particular satellite. If the satellite were owned by the French, they could program propagandistic material, which might not be welcomed by governments in Germany or Spain or Poland. The possibility even exists that non-European countries such as Russia or the United States could own satellites over Europe (or any other part of the earth) and program as they see fit. Propagandistic programming is common now through shortwave and through limited overlap of regular television signals from one country to another, but the political problem of satellite transmission is larger than any countries have tackled to date.

These international problems are not going unheeded. Many conferences have been held on the subject, most notably through the United Nations Agency, the International Telecommunication Union (ITU), which oversees periodic World Administrative Radio Conferences. COMSAT, INTELSAT, the Organization of American States, the FCC, the BBC, and the European Space Agency (an alliance of eleven European countries and associated members) are among the organizations studying the international satellite situation. They hold meetings and do make some progress, but frequently they reach stalemates on issues. In the meantime the countries, primarily the United States, with the means to launch satellites, continue to do so.[19]

### Backyard Satellite Reception

Much closer to home, the controversy arises over the right of individuals to receive, for free, programming intended primarily for other entities, such as cable TV.

People who have satellite dishes argue that they paid to receive the programming when they purchased the dish, and say they are entitled to whatever is in the air space above their homes. In addition, many of them are in remote areas that can not receive television by other means. The manufacturers of satellite dishes agree and state that satellite programming suppliers are trying to curtail the growth of the dish business for the benefit of cable TV.

On the other side, satellite programming providers argue that receiving such signals is piracy and a violation of copyright. They feel totally justified in scrambling their signals, even though it is of significant additional cost to them.

The scrambling of signals may lead to the use of illegal decoders which will simply heighten the controversy. In addition, because there are so many signals that can be picked up by satellite dish owners, they may simply ignore the signals that are scrambled and watch unscrambled signals.

Program suppliers may set up a mechanism for charging people with backyard dishes, but this will involve marketing on an individual basis which could prove costly. Several companies are considering taking a number of scrambled signals and selling them as a package to backyard dish owners. This may or may not work. At best, such a plan is several years down the road because the scrambling has not been put into total effect as yet.[20]

## Satellites and the News

Because satellites can provide worldwide instant communication, they have greatly increased the speed with which worldwide events can be shown to the public. Less than 200 years ago the Battle of New Orleans was fought a week after the War of 1812 had ended because the news of the war's end took that much time to reach Louisiana. Today, largely due to satellites, the slightest rift between countries can be reported, analyzed, and even blown out of proportion within a matter of minutes.

And sometimes that is what happens. Events, which at one time would have been of little significance, are reported because they can be. In the race for viewers, news organizations try to out-scoop each other with pictures of minor squabbles in far-off lands. Sometimes these squabbles might disappear of their own accord if they were not so highly reported, and if the people involved were not forced to make statements to the press that put their views very firmly on record. This eliminates any graceful possibilities for backing down on issues and can exaggerate the extent of the disagreement.

The ability to communicate instantaneously can lead to another flaw, inaccuracies. Sometimes in their rush to be first with an item, reporters do not take the time to check out all the facts. In the days when communication was much slower, inaccuracies could be caught before they had traveled very far.

Eliminating satellites will not eliminate these journalistic shortcomings. Only responsible reporting can do that, but unfortunately, satellites often help emphasize rather than correct the flaws of the news gathering process.

## Teleconferencing Controversies

As the frequency of teleconferencing rises, and the cost lowers, some interesting sociological phenomena can occur. If meetings are conducted by having participants address each other with cameras, dishes, and satellites as an intermediary, will the human part of the communication disintegrate? Can either hard negotiating or camaraderie exist between people or groups who are not physically close? If conventions are held via satellite, what will happen to the very valuable informal exchanges that occur outside the formal meeting room sessions?

A full-blown growth of teleconferencing could greatly diminish business for the transportation industry, especially the already hard-plagued airline companies. If business trips are significantly reduced, the airline companies will need to undergo major restructuring.

The hotel industry could also be hit by increased teleconferencing. If conventions are held through multiple satellite feeds, people can attend the convention at a place near enough to their home that they do not need to stay overnight. Obviously, this will reduce hotel business. To counter this, several hotel chains have equipped themselves to become videoconferencing centers so that they will be chosen as the local link, and will, therefore, profit at least from the local business.[21]

Satellites can change the social fabric of the future more than they have in the past. The technology, although it is one of the oldest of the "new technologies," is still in an infant stage. It has the capability of changing the world into what Marshall McLuhan prophesized as a "global village."

# computers

# 4

## Description

Computers are intrinsic to many facets of the new television technologies. They count responses of interactive services; they turn videocassette recorders on at the proper time to record off-air programming; they control satellites; they provide endless amounts of information within videotext and teletext services; they aid in scrambling and unscrambling signals for subscription TV; they create the graphics and animation seen on many of the systems; and they even bill cable TV and subscription TV subscribers.

Without computer technology the new television technologies could not exist in their present forms, and many of them could not exist at all. Computers are as essential to television as they are to many other facets of society.

### Basic Operations

Basically a computer is a device which processes and manipulates data very rapidly. In order to do this, it stores information in a memory; it receives input from various sources; and it outputs the results of its processing.

Various devices attach to the computer to accomplish all this. Tapes, disks, keyboards, and punched cards are among the devices which give input to the computer. This input is then stored in memory and processed on tiny chips which constitute the actual computer. The results of the processing can be sent to such devices as tapes, disks, cathode ray tubes, and printers.

### Processing

The part of the computer that undertakes the processing and manipulation of data is the central processing unit (CPU). This CPU usually consists of a small silicon chip which is not even visible in a computer set-up.

The CPU receives instructions from software programs written to undertake various tasks. For example, software has been written to enable computers to make financial predictions based on information concerning past and present finances. Software development has enabled computers to undertake an almost unlimited number of processes. They can update information such as the amount of money in a checking account. They can sort by one or many factors such as social security number, age, and geographic location. They can select certain elements such as all students with a grade point average of 4.0. They can respond to moves someone makes playing a game and create new moves for the game. They can summarize data such as total sales for each salesperson. They can teach math by presenting supplementary material when a student does not solve a problem correctly. They can classify data into certain categories such as food stuffs, clothes, and automotive supplies.[1]

## Memory

All of the processing which occurs in the CPU is directed by programs which are stored in the computer's memory usually consisting of one or several chips. Data to be processed is also stored within the memory chips.

There are two main methods of storing information in the computer, Read Only Memory (ROM) and Random Access Memory (RAM). With ROM, information is recorded in the memory of the computer when it is manufactured and this information can not be altered by the computer user. The information can be read and used but can not be changed. It usually relates to the basic operation of the computer.

Random Access Memory is used for all information which might be changed. It can be written into and read from storage. RAM is the primary form of storage for most computer applications and is certainly the type of storage with which the average user of a computer is most familiar. Any time words, numbers, charts, or other data are stored in a computer RAM is utilized.

Information is stored in a computer's memory in a form that is different from what human beings usually encounter. A computer can not recognize regular letters and numbers. It can only recognize an on or off state—it senses the presence or absence of electronic impulses. Theoretically, an "a" could be on and a "b" could be off, but that would be the total of what a computer could recognize. To solve this problem, codes have been developed that deal with bits and bytes. The bit is a single unit of storage that can assume either the on or off value. Eight bits make up a byte or a string of possible ons and offs. Numbers and letters are represented by varying the bits that are off and on within a particular byte. For example, for the number one, bits 1, 2, 3, 4, 5, 6, and 7 are off and bit 8 is on. For the number two, bits 1, 2, 3, 4, 5, 6, and 8 are off and bit 7 is on. The eight-bit bytes allow for 256 possible combinations.

Computers are manufactured with different amounts of memory. Some store as few as four thousand bytes and are known as 4K computers (K stands for kilo which means thousand). Others may have 16 million bytes, usually referred to as 16 megabytes.[2]

## Input-Output Devices

A variety of devices are used to place data into a computer so that it can be processed. Some of these same devices serve as outputs to display the results of the processing undertaken in the CPU.

Two of the most common input-output devices are magnetic tape and magnetic disks. Magnetic tape is similar to audio and video tape. Data is stored by recording electronic impulses on magnetic oxide. Magnetic disks record similarly to magnetic tape but in a circular rather than linear pattern. Tapes are inserted into tape readers and disks are inserted into disk drives. Both the readers and the drives can be connected to the central processing unit. The CPU can then call up information from the disks or tape to process it.

Once the data has been processed, the results can be sent from the CPU to the disk drive or tape reader and placed on the disk or tape. In this manner the disks or tape become output devices as well as input devices.

Another common input device is a keyboard. The computer keyboard usually resembles a typewriter keyboard with a few additional keys used to move and correct the characters typed by the operator. Signals from the keyboard are sent directly to the CPU for processing. A keyboard does not serve as an output device because it can not receive or store data.

Electronic impulses sent to the CPU from tape, disks, or a keyboard can not be seen by the human eye. Some sort of display must be incorporated with these devices to allow the operator to visualize what is happening in the input and output operations. This display is usually a cathode ray tube (CRT). It is a screen similar to a TV screen but with higher resolution than a TV set so that small letters and numbers can be seen easily.

Material typed on the keyboard is routed through the CPU and then displayed on the CRT. Changes can be made to this material through software stored in the computer memory and used by the CPU. For example, if an operator wants to move a column of numbers to the left, a few keystrokes will instruct the CPU to move the numbers and the new position of the numbers will be displayed on the CRT.

Material typed on the keyboard and sent to the CPU can then be sent to disks or tape for storage. At some later time, this material can be sent from the disk or tape to the CPU and then to the CRT. In this manner, disks and tape serve as input, output, and storage.

A common output device for computer systems is a printer. Once the material has been processed and is in a final form, it can be printed onto paper. Many varieties of printers are available for computer output. Some are solid character printers that operate on the same principle as a typewriter. Others are dot matrix printers which do not print whole letters or numbers as the solid character printers do. Instead, each letter or number is created by a series of dots. Solid character printers produce written material that is easier to read and more professional looking, but dot matrix printers are faster and cheaper and capable of printing charts, headlines, and other graphics.

The cathode ray tube is another output device. Once material has been processed, it can be viewed on the CRT. Often it is not necessary to print this information. It can simply be viewed and noted so that further work can be undertaken.

**Figure 4.1.** Computer configuration.

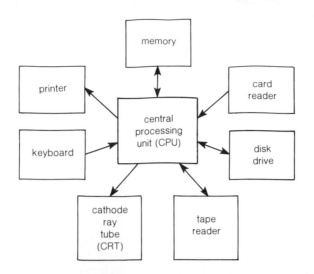

There are many other input and output devices tied to various types of computers. Older computers use punched cards for input. These contain holes which represent data and can be read by a card reader into main computer storage.

Sometimes the output of a computer is placed on microfilm. Microfilm holds microscopic images on film that is easily stored. In order to read the material, a person must use a microfilm viewer which magnifies the image.

The buttons used to program a VCR to record at a particular time are input devices. So is the small keypad comprised of numbers 0 through 9 used to retrieve teletext information. Some input devices have been developed to read from original material by recognizing words or other figures on a printed page. For example, some devices can read bank checks and input the information to the CPU. The optical bar code which now appears on groceries and other products is read directly into computers.

Currently under development are input and output devices which utilize audio. For these, a person does not need to type anything. The operator can simply talk in plain English and the words will be translated for the computer. Likewise, a computer can simulate a voice response. Some telephone directory assistance systems are now using this. After asking an operator for a particular number, the caller hears a computer-generated voice giving the phone number. Voice input and output are still in the experimental stages, but eventually people may be able to talk to computers and receive verbal responses.

The number and variety of input and output devices needed for a computer set-up differs for different individuals and companies depending on how they use the computer.[3]

## Computer Interconnections

Frequently computers, or parts of computer systems, are tied together in various configurations. One of these is called time-sharing. Keyboards and cathode ray tubes (commonly referred to as terminals) will be placed in various locations for the inputting of data. They will each be wired to a central computer (often called the mainframe) which is used for processing. Because inputting usually takes much longer than processing, quite a few terminals can use the same computer without overloading it. Often these terminals also share one printer which is used only when the input has been completed and corrected and is ready for final output.

Numerous computers can also be tied together so that they can send and/or receive information from each other over phone lines. This is accomplished through a modem (short for modulator/demodulator), a device which connects a computer to a telephone. It encodes the computer output into audible tones so that it can be sent over telephone lines to another telephone which is connected to another modem. This second modem translates the tones back into computer form so that it can be manipulated or displayed by the second computer. By using this form of computer communication system, someone can compose a letter or graph on a computer and send it to a friend or business associate in another part of the country.

A large number of computer data banks have been established which give information to subscribers. Two of the largest are CompuServe and The Source. Each of these contains news, stock market reports, recipes, movie reviews, and other information of use to home users and businesses. Anyone with a computer who pays a fee can receive this information by dialing special numbers on a telephone connected to a modem. Some of these data banks are linked with the new television technologies, particularly videotext.[4]

## Uses

Computers are becoming omnipresent in all facets of society. They are built into watches, automobiles, and numerous household appliances. Airline tickets, theater tickets, tickets to sporting events and many other activities are booked by using computers. Inventories, sales, profit and loss figures, and many other aspects of business are tracked through computers. The medical industry uses them for diagnosis and monitoring.

Word processing has become a major use for computers, making the typewriter almost obsolete in many businesses and homes. Long documents can be typed, displayed on CRTs and stored on disks. Mistakes can be corrected with a simple keystroke, eliminating messy erasers and correction fluid. Documents can easily be right hand justified, and words can be boldfaced, underlined, and centered with ease. Margins can be changed and pages can be reformatted after something has been typed. Paragraphs can be inserted and moved with equal ease. All of this makes writing and editing material much easier than was possible before computers.[5]

One element of the new television technologies that makes great use of computers is video games. The early home games were self contained with each game in its own computer based housing. Later games consisted of programmable housing equipped with a computer and different memory cartridges which plugged into the programmable housing for different games. Cartridges were then developed to make use of the same housing.

The computer games simulate a normal broadcast signal by internally creating their own pictures. Each game divides the television screen into approximately 2500 squares, which it can turn on or off or turn to a particular color. An array of these squares creates a playfield. As a player makes various moves they are relayed to the computer. The computer then calculates its counter moves and revises its playfield. In this manner, the game is a battle against a computer, albeit a benign friendly computer.[6]

These uses, and many more, have made society more dependent upon the computer than most people realize.

## History

Computers have a very short history, spanning only several decades. Their use within the new television technologies is even shorter. The history has been interesting—replete with fast-paced technical breakthroughs, competition, and economic growth and decline.

The history of computers can be broken down into four generations—the first dependent upon the vacuum tube, the second brought about by the invention of the transistor, the third the result of the silicon chip, and the fourth due to very large scale integration (VLSI).

### The First Generation

The development of the first computer is usually attributed to Dr. John V. Atanasoff, a math professor at Iowa State College. In the late 1930s he and an assistant, Clifford Berry, designed and built an electronic digital computer which they named the Atanasoff-Berry Computer or ABC. Their purpose in designing it was to help twenty Iowa State master's and doctoral candidates compute mathematical operations. It utilized bulky vacuum tubes for the storage and computation of data.

Several years later two other people with computer interests, Dr. John W. Mauchly and J. Presper Eckert, Jr., heard about the ABC and communicated with Atanasoff regarding it. When World War II broke out, Mauchly and Eckert learned that the army had a need to calculate ballistic tables to produce trajectories for bombing. These calculations took about fifteen minutes using the mechanical calculators then available. This was much too slow to accommodate the trajectories which took only one minute. Mauchly and Eckert submitted a proposal to the army for a machine which would make the calculations in thirty seconds and were awarded a contract to develop it. In 1946, after totalling expenses of $400,000, Mauchly and Eckert completed the computer they called ENIAC (Electronic Numerical Integrator and Computer.) It contained 18,000 vacuum tubes, weighed thirty tons, and could multiply two numbers in about 3/1000th of a second. After the war ENIAC was used for purposes such as weather prediction and random number studies.

One of the problems with ENIAC was that for each different purpose, all the wires had to be moved around. In essence, the only way to program the machine was to rewire it. In the 1940s, Dr. John Von Neumann of the University of Pennsylvania developed the stored program concept wherein the instructions for the computer were stored in the computer, itself, and could be changed easily by inserting new instructions. This is a basic concept which still exists in computers today.

Meanwhile, Mauchly and Eckert had formed a company to design and build computers for government and industry. This company was purchased by Remington-Rand, making it the first major company to enter the computer business. Mauchly and Eckert, working for Remington-Rand, developed the Universal Automatic Computer called UNIVAC I and sold it to the U.S. government to use for the 1950 census.

Up until this time the general public had little or no knowledge of the development of computers. Television, however, changed this. In 1952 CBS used UNIVAC I for its election coverage. With only five percent of the vote tallied, the computer predicted that Dwight Eisenhower would defeat Adlai Stevenson for president.

During the 1950s, IBM, the leading company in producing electro-mechanical business machines using punched cards, decided to enter the computer business. IBM and Remington-Rand engaged in keen competition for several years with IBM eventually winning out. This was largely due to the introduction of the IBM 650 which was designed for business applications. IBM's dominance in the business machine field gave them a great marketing advantage because businesses saw the 650 as the next step up from the punched-card accounting machines they had already purchased or leased from IBM.

One of the difficulties shared by all the early computers was that the programs needed to make them operate were very difficult to write. Programmers had to write instructions in a language that computer hardware understood. In other words, they had to adapt to the fact that the computer only understood the concept of on and off. Writing such computer programs was laborious and error-prone.

Several IBM employees developed a language called FORTRAN that allowed programmers to write in mathematical notations which the computer, itself, would then translate into its machine language. In short succession, many other programming languages were developed and the problems involved with programming decreased.[7]

## The Second Generation
The second generation of computers was spawned by the invention of the transistor. Scientists at Bell Laboratories invented the transistor in 1947 but it did not become an important element in computers until about ten years later.

The transistor served the same function as tubes but was much smaller, cooler, more reliable, cheaper and faster. Companies with smaller budgets, and smaller rooms with less air conditioning could accommodate these computers.

The first transistorized computer (TRADIC) was built by Bell Laboratories in 1954 and contained about 800 transistors. IN 1958, IBM brought out two computers which used only transistors. These machines and those built by different companies ushered in the second generation of computers.

The high sales of these second generation computers further entrenched computers into American society, and especially the business world.[8]

## The Third Generation
The third generation of computers was ushered in by IBM's 1964 announcement of its System/360 family of computers. System/360 computers used silicon chips rather than transistors for the storage and processing of data. For chip technology, transistors and related circuits were drawn according to specifications and then photographed and reduced. They were then etched onto a thin wafer of silicon about the size of a fingernail. These chips reduced computer size and cost and increased speed and reliability.

System/360 computers were designed for all types of processing. In the past, machines that undertook scientific computations were manufactured in a different manner than computers that were used for business purposes. All the System/360 computers could be used for both scientific and business purposes.

The initial System/360 computers had severe software problems which IBM programmers had to work long and hard to overcome. Eventually these problems were resolved and the 360 became the largest selling of any machines up to that time.

In general, the 1960s saw a booming computer industry. Computer systems were sold and rented at a dizzying pace. The demand for programmers far outstripped the supply, and people with any type of programming knowledge could demand high salaries. Both hardware and software companies sprung up like mushrooms.[9]

## The Fourth Generation
The fourth generation of computers began in the 1970s when the number of electronic components which could be placed on a chip multiplied significantly. A chip of the third generation could contain approximately 1000 circuit elements. The chips of the fourth generation using what was referred to as very large scale integration (VLSI) could contain approximately 15,000 circuit elements. One computer company, Intel Corporation, designed a whole computer on a chip and called it a microprocessor. This one chip almost matched the power of the original 18,000 vacuum tube ENIAC computer.

Both tape and disks became well developed and accepted during the 1970s, and in many instances replaced the punched cards which had been used for years as input devices.

All of this further reduced the size and cost of computers and led to the development of personal computers which could fit on a desk top. By 1975 electronic hobby magazines were advertising computers which could be assembled from a kit costing $395. The Apple computer was developed in the garage of two computer hobbyists and became a huge success in the home computer market. Many other companies entered the personal computer business—Radio Shack in 1977, and IBM in 1981.

Overall, the late 1970s and the early 1980s did not yield the booming times for the computer industry that had been the hallmark of the 1960s. Part of this was due to the fact that too many companies had entered the field and a shakeout period was necessary. This shakeout period saw the reorganization and death of many hardware and software companies. IBM remained the dominant company in most areas of computer sales, making it hard for smaller companies to compete.[10]

Computers had arrived to stay, however, and their overall impact on society was acknowledged. Just about every aspect of life embraced them in some form, and the new television technologies were no exception. They became essential elements in television equipment and they became the primary means through which consumers interacted with various TV technologies.

## Video Games

One of the new television technologies very closely tied to computers has been video games. The history of this phenomenon has been replete with eccentricity and financial cliff hanging. In many ways this history typifies what happened in the computer industry, exaggerated a bit perhaps because video games tie together two volatile industries, computers and television.

Atari, the oldest, largest, and most successful of the video game companies was founded in 1972 by Nolan Bushnell, a young, egotistical but likeable engineer who, as a student at the University of Utah, had played games with computers and worked at an amusement park during vacations. Bushnell named his company Atari after a term in a Japanese game which politely warns an opponent of imminent annihilation. Bushnell's first invention, a game called Computer Space, was not successful, but then he hit upon the idea for Pong, an electronic version of table tennis. This was a very simple game that revolutionized the coin-operated game business then supported primarily by the pinball machine.

Bushnell did not profit greatly from this rage because a horde of competitors snatched most of the Pong business. Only about 1/10th of the Pong machines were made by Atari. Other games developed by Bushnell and friends he had brought into the company never caught on, and the company perched on the brink of financial disaster.

Meanwhile Magnavox introduced the first home video game called Odyssey. It failed, but Atari took the bait and developed a home version of Pong which sold out almost instantly. The small company could not manufacture enough games quickly enough to meet demand, so Sears lent financial help to allow Atari to increase production. It became obvious to Bushnell and others at this point that if Atari were to grow and maintain inventory, it would need to go public or sell to a large company which could invest heavily.

After brief romances with MCA and Disney, Atari was bought in 1976 by Warner Communications, primarily because of the enthusiasm of one of Warner's top executives, Emanuel Gerard. But even Warner's financial clout could not move Atari off the ground. Consumers seemed weary of dedicated video games which, except for Pong, had never really been a hit. So Atari's engineers developed programmable cartridge model video games and introduced them in 1977. Similar games were developed by RCA, and Fairchild. None of the games caught on.

Meanwhile, intrigue captured the executives ranks of Atari. Bushnell and Gerard had several shouting matches resulting in Bushnell's ouster from Atari in 1978. The new chief executive, Raymond Kassar, was a straight-laced executive who totally turned around the informal casual atmosphere Bushnell had created within the company. He mandated that people wear suits and ties instead of T-shirts, checked to see that everyone reported to work promptly at 8:00 A.M., and antagonized the engineers by referring to them as prima donnas. As a result, most of the people who had worked for Atari under Bushnell quit.

Kassar, either because of his disciplined approach or because the time was ripe, produced results. He hired a top rated marketing staff and launched a $6 million advertising campaign, and in 1979 retailers snapped up all the video games Atari could manufacture. In fact the video game industry began to bring in more revenue than records, radio, TV, or movies. For awhile Atari accounted for over half of Warner's earnings, and games such as Space Invaders, Asteroids, and Pac-Man became instant celebrities and temporary national past times. In 1981, Atari sold $2 billion worth of video games.[11]

This success brought competition from many small new companies hoping to cash in quickly on the craze. Most of these failed almost immediately because they were undercapitalized. In addition, the large toy company, Mattel, entered the market with Intellivision. Technological problems and management upheavals within Mattel prevented it from placing its product on the market until late in 1979 and by that time Atari had a large number of its consoles already placed in homes. Intellivision sold somewhat successfully but never overtook Atari.

Then, as quickly as the video game craze had surged, it ended. By 1982, interest in home video games had subsided, and Warner and Mattel stocks plummeted as both companies admitted that their 1982 sales predictions would never be met.[12] In 1984 Warner sold the home video game portion of Atari to Jack Tramiel, the founder of Commodore International, a computer company which had developed a successful line of home computers. Warner reported a $425 million loss as a result of this sale. Tramiel is trying to revive the video game craze through new game concepts. Given the topsy-turvy history of the video game business, his success is unpredictable.[13] At present video games located in arcades continue to draw moderately well but the home video game market is languishing at best.

## Issues

Computers have created unrest and distrust in society. Their silent encroachment bothers many people who do not feel they can comprehend the technology. People fear that computers will take away their jobs, invade their privacy, destroy interpersonal relations and create artificial dependencies. Within the computer industry itself there is concern over economic downswing and over what has come to be called computer crime.

### Job Replacement

Computers can replace people. They can make calculations more quickly than rooms full of accountants and they can be programmed to undertake the same tasks that machinists and draftspeople usually do. They can teach, draw, and create music.

On the other hand, they create jobs, primarily in the area of computer programming. Not every accountant, machinist, or draftsperson wants to become a computer programmer, however. People who have been replaced by computers are often bitter because they must be retrained for jobs which are not always appealing to them.

Computer advocates argue that the jobs which computers handle are the tedious, repetitious ones. By taking over these functions, computers free human beings for more meaningful, creative endeavors.

Computers can also make people more productive. A writer using a word processor can write much faster than one using a paper and pencil. A teacher who utilizes computer-aided instruction can design individualized learning programs to teach students faster and more thoroughly. The computer will not replace the teacher or the writer; it simply makes their jobs easier.

Many people who have not been replaced by computers have, nevertheless, had to undergo changes in their workstyle to accommodate computers. Supermarket clerks checkout groceries differently after computers are installed. Librarians have had to be retrained in methods of handling books. Retraining to accommodate computerization often causes trauma in the workplace. Many people who had been doing something in a comfortable manner for years do not see the need for change. The old method worked well. In fact, in many instances it worked better than the new computerized method, especially when the computer system is first installed and has not had all the bugs worked out of it. Resistance and resentment often build up when computers are introduced.

There are people, however, who appreciate the challenge of the computer. They have often been in the same job for years and are beginning to feel stale. Learning something new gives them a brighter outlook on their work.

Despite the problems associated with job displacement and retraining, computers have made tremendous inroads into many areas of the workplace. In fact, in some industries companies are forced to become computerized in order to remain competitive.[14]

## Privacy

An individual's privacy is something that is cherished in the American society. The home is a sacred place which can not be searched without a warrant. Our forefathers wrote protective measures into the Constitution to protect privacy, but they never anticipated computers.

Now there are large data banks which contain information about people which, if put in the wrong hands, could be very damaging. Potential kidnappers who can determine people's net worth could make outlandish demands. Potential robbers could learn about valuables people keep in their houses. On another level, salespeople could become bothersome after learning of past purchases people have made.

Security measures are taken to protect information in data banks, but no computer is totally secure. If someone really wants to obtain information, he or she can.

Information in data banks does help with many innocent statistical calculations. By going to the right data file, someone can calculate such information as the average income in Los Angeles versus the average income in St. Louis, the increase or decrease in divorce rates or test scores of school children, or the projected number of people between the ages of 18 and 24 in the United States in the year 2000. This kind of information is useful for individual and social planning.

The information in data banks is somewhat secure in that it is scattered in different locations. A plan was proposed in the early 1970s for a national data bank that would contain a great deal of information on each citizen of the U.S. This would include name, address, medical history, education, criminal records, credit rating, income tax data, and marital status. The idea was abandoned because people felt this would lead to privacy violations. Much of this information is in computers; it is just in separate files around the country.

Similarly, a plan was proposed for an electronic funds transfer system which would create a cashless society. Paychecks would be automatically deposited in an individual's fund. Any purchases that a person made would be deducted from the fund. This would have solved the problem of bad checks and poor credit purchases, but it would have created a society where all purchases made by everyone would be known. This idea was rejected, again on privacy grounds.

Closely related to privacy is the issue of errors. Someone who receives a credit rejection because incorrect information was placed in the computer often does not even know of this error. People have been arrested because a check run conducted when they were stopped for speeding revealed a criminal record that was incorrect. Needless to say, these people harbor resentment which they often direct toward the computer.

Bills have been proposed at the state and national level to protect people from computer invasion. This legislation is difficult to write, however, because computer technology changes so rapidly.

Invasion of privacy is not a large problem, but its potential is large. As the amount of information that computers gather increases, safeguards must be found to protect individuals.[15]

## Decline of Interpersonal Contact

Computers are by nature impersonal. They are not people. When a computerized voice calls to tell someone that a catalogue order is in, there is no way for that person to ask how late the store is open. If someone sits at a home computer terminal to order clothes or tickets, they do not interact with other people in stores or offices. A society were computers talk to computers is far less personal than a society where people talk to people.

Computers can also program people to respond in ways that may be impersonal. The argument has been put forth that computer games teach people to respond in certain ways in order to win. By the same inference, people could be trained to learn in certain ways and to gather data by certain patterns. This could destroy individuality and lead people to respond to everyone in a similar manner.

Such high degrees of impersonality are not likely, given the somewhat naturally gregarious nature of most human beings, but computers can certainly diminish interpersonal contact.

## Dependency

As each day passes, the world becomes more dependent upon computers. They cook our food, dry our clothes, operate our automobiles, and in many ways schedule our lives. Most of the time people do not even realize the impact computers are having upon their daily activities. If all computers suddenly stopped functioning, life, as we now know it, would have to change radically.

It may not be healthy for individuals and society to be as dependent upon computers as they presently are. At the very least, manual backup systems should be maintained. But they usually are not. Companies that convert to computer operations rarely continue duplicating these operations in whatever their "old fashioned" way was. People who buy home computers often discard their manual and electric typewriters.

Anything that is depended upon can become powerful, and, although it is human beings that determine exactly what computers should and should not do, the ease and convenience of computers may lead people to be dependent to a greater degree than is healthy.[16]

## Economic Downswings

The computer industry is by no means down and out, but quite a few companies in computer-related businesses have run into financial difficulties. Some of them deserved their problems. They saw computers as a way to make a fast buck and set up business without vision or capital. These companies would have failed regardless of the line of work they undertook.

But the computer industry was also overly touted. Predictions regarding the number of computers which would be sold or the number of software programs which would be purchased were often overly optimistic. The growth of the personal computer business, for example, did not meet expectations. But then, the expectations may have been unrealistically high.

The abilities of computers were also over touted. Software has frequently been unreliable, to the extent that some companies have turned away from computers and back to other ways of handling business. The expense of computers has not always outweighed their value. This is especially true if a computer performs a particular task that is non-repetitive or if the amount of data involved is small. The amount of time it takes to program a computer to do something once is often greater than the time it takes to do the job. Many companies with low-volume business found it was easier to make calculations by hand than to set up the computer for the operation. Similarly, individuals who bought computers so they could balance their checkbooks often found that it took more time to input the financial data into the computer than it took to balance using a calculator. As a result, some people became disenchanted with computers and did not purchase them or their related software at the rate predicted.

Regardless of the reason, the computer industry has experienced a fallout period which may prove to be healthy in the long run, but causes concern to those directly affected by economic and personnel cutbacks.[17]

## Computer Crime

Computer crime is still a largely undefined area. Someone who breaks into a computer file to find out how much money his neighbor has in the bank is probably clearly guilty of computer crime. But less onerous is the hacker who breaks in just for the fun of it.

Closely related to computer crime is computer ethics. People frequently make copies of software programs and give them to friends. In some instances this is a violation of copyright, but it is a common practice. Sometimes people use a computer at work to do calculations for their own personal use, or they give out secret passwords to friends who can use the computer during off-hours. All of these are gray areas which need to be discussed, if not resolved.

Employees who have access to computers can commit acts of dubious criminal or ethical nature. In one instance a man who was in charge of monitoring computer crime used the computer to check on his wife's charity contributions. Bank employees have been known to transfer funds to their own accounts to cover short term debts. Although they transfer funds back several days later without anyone being the wiser, they have acted in an unauthorized manner.

Some states and the federal government are enacting computer crime laws, but the groundrules for and interpretations of these laws are not yet well developed. Usually such laws outlaw the use of computers for fraud or extortion and forbid the alteration of data. After numerous cases have been tried under these laws, their interpretation may become more refined.[18]

## Video Games

During their heyday, video games were severely criticized. Doctors identified a new ailment called "Space Invaders Wrist," a strain of ligaments and joints caused by playing Space Invaders for long periods of time. Many children, and some adults, became essentially addicted to the games, playing them both at home and in arcades for hours on end—hours that could have been spent doing homework or participating in fresh-air athletics.

Parents who at one time admonished their children for watching too much TV began to scold them for playing too much TV. Psychologists were worried that the stress which the games placed on power, aggression, and manipulation would have a bad influence on children.

On the other hand, there were those who defended the games because they helped develop motor skill and accuracy on the part of the players. They were not idle chance games. Skill was required. In fact, many of the children who mastered the games developed self confidence. Because children were fascinated with the games, they were more receptive to educational methods that paralleled the games. Many young children became enraptured with computers, and some sociologists attributed this, at least in part, to the fact that they had been introduced to computers through video games.

Now that the success of the games has waned, the criticism, as well as the praise, has waned. Anything in excess can be harmful, and for awhile some children did indulge in video games to excess. Now these games seem to present a moderate past time undertaken primarily in arcades and not in the home.[19]

# part 3
## new technologies

# cable tv

# 5

## Description

Cable TV is basically a method of receiving video information through wires rather than over the air, as is the case with regular broadcast TV. From a technical point of view there are advantages and disadvantages to each system.

### Cable Compared to Broadcast

For broadcast TV, an antenna is placed at a high location, and a signal is sent out in a particular frequency range. Any TV set manufactured to be able to tune to that particular frequency range can pick up the signal, as long as the TV set's antenna is in "line-of-sight" with the broadcast antenna. This means there must be no major obstructions between the set antenna and the broadcast antenna, or the set antenna must be able to pick up a clean reflection of the signal from a hill or building. If such is not the case, then TV reception is poor or nonexistent.

For cable TV, however, the TV set is not dependent on line-of-sight with an antenna. Video signals are delivered on wires instead of through the air. The cable TV facility receives the signals from broadcast antennas and from satellites and then places these signals on wires, which are buried underground or strung on telephone poles. The wires pass every home in a particular neighborhood, and those people who wish cable TV service pay to have a wire brought into their home and attached to their TV set.

Another difference between cable TV and broadcast TV is that cable TV has the capability of being two-way while broadcast TV can be only one-way. Because signals are brought into the home on wires, signals can also be taken out of the home by wire and sent back to some specific place such as the cable main office or a bank of computers. Obviously, people cannot have transmitters in their homes to send messages back to broadcast TV stations.

Still another difference between cable TV and broadcast TV is that a cable TV service is concerned with many channels while a particular broadcast service will be concerned only with one channel. Cable TV provides a viewer with numerous channels of programming while a particular network or particular station provides only one channel.

Many of the early cable systems supplied three channels, one for an ABC station, one for a CBS station, and one for an NBC station. As television and its resulting technology grew, cable systems provided as many as twelve channels of programming. The cable system could use each channel from two through thirteen because its signals were on wires that were not subject to the same interference that makes it impossible to use all twelve channels of broadcast TV in a particular area. When video information travels through the airwaves the frequencies assigned to each station can interfere with each other easily. However, when video information is placed on wire, the signals can be insulated from each other so that all the different frequency ranges can be utilized.

Different cable systems placed different programming on these twelve channels but usually it consisted of all the local VHF stations plus all the local UHF stations converted to VHF space on the dial. If this was not a large number of stations, then the cable system would bring in stations from near-by communities or perhaps program some material itself. As technology continued to improve, equipment became available that enabled cable TV to include twenty channels, then thirty-seven, then fifty-four, then upwards of 108. At relatively the same time satellite distribution of program material became a reality, and more programming became available to fill the multitude of channels.

Although it may appear that cable definitely has the advantage because it does not need to be line-of-sight and because it can provide more channels, there are certainly disadvantages to the cable TV technology. Broadcasters can simply place an antenna on a hill and send out their signal. They do not need to be concerned with the reception process. Anyone who can place an antenna high enough to be in line-of-sight with their signal can buy a TV set and watch the programming.

Such is not the case with cable TV. The process of stringing the cable and providing all the equipment necessary to bring signals to the viewer is time-consuming and expensive.

## Headends, Hubs, Cable, and Converters

First a cable system must build a headend. This is a site that receives all the signals which will later be sent to the cable TV subscribers. This headend will consist of at least one satellite ground station dish. The headend will also consist of antennas to receive the regular off-air AM and FM radio stations and UHF and VHF TV stations. It may consist of microwave dishes for receiving more distant stations or even wire terminals for receiving information generally sent over wire, such as news bulletins from the wire services. Audio only services such as syndicated music or radio programming from a local college can come through wire or satellite transmission. If the cable system originates its own programming, then playback tape machines may be part of the headend. All of the signals come in to the headend and undergo processing so that they are placed on particular channels to go to the consumer.

At this point, any material can be placed on any channel. The cable company can decide to place Home Box Office on channel 4, ESPN on channel 20, the weather information on channel 12. It can even decide to place a local channel, which is known as channel 4, on channel 21. This is not usually done, however.

Cable companies generally keep all VHF signals on the channels people are used to, but they do frequently place UHF channels on numbers other than their broadcast channel number. Audio signals can be placed on empty FM channels and delivered to subscribers' radios.

After the material gathered at the headend is processed onto particular channels, it is placed on coaxial cable to be sent through the wired system to homes. Sometimes the area that the cable company serves is too large for cable alone to do the delivery job. As the signals go through wires, they lose some of their strength and need to be amplified. Amplifiers along the way can do the job for awhile, but eventually the signal becomes so weak that simple amplification is not adequate.

If such is the case, then the cable system must construct hubs. Hubs are essentially microwave receive sites to which the signals are sent so that they can be distributed by wire, thereby eliminating several miles of coaxial cable and reducing the number of amplifiers. A headend will process all the signals and then send them through the air to hub microwave dishes. This necessitates buying a sending microwave dish and a receiving microwave dish for each hub. When the signal arrives at the hub, it is still fairly strong but can be further amplified.

The signals received at the hub are then placed on coaxial cable and distributed to an area of homes located near the hub. In this type of configuration, headends usually also act as hubs in that signals are distributed by wire from the headend to homes near it.

If the cable is strung on telephone poles, the cable TV company must gain permission to use the poles and pay the phone company for this use. Because the phone company often has to move some of its wires around in order to make room for the cable and because historically there is no great love between the phone companies and the cable TV companies, the process of obtaining room on the poles is often tedious and drawn-out. If the coaxial cable is to be placed underground, then the cable TV company must secure all the necessary permits to dig up city streets and right-of-ways. This, to, can be a drawn-out process with many unforseen impediments. In addition, placing cable underground is much more expensive than stringing it on phone poles.

The cable company maps out the area it serves so that cable will run close to every building. If a homeowner wishes to subscribe to the service, the cable company taps a wire off the main line passing the home and connects that wire to the subscriber's TV set. Dealing with people in apartments or condominiums is more complex. Cable systems try to convince landlords to buy the service for all tenants, but that approach has not been highly successful. Sometimes apartment owners want financial remuneration for themselves in order to take the service. Taking lines to individual tenants often creates an unsightly group of wires around the apartment complex. In addition, access to attics and other places where cable needs to be strung can be difficult.

Once the wire reaches the TV set, there is still more equipment needed. Cable TV systems that have more than twelve channels of programming cannot operate

**Figure 5.1.** A cable TV headend with
a microwave tower and satellite dish.
Photo courtesy of Valley Cable TV.

**Figure 5.2.** This hypothetical franchise area shows how the headend and hubs relate to each other.

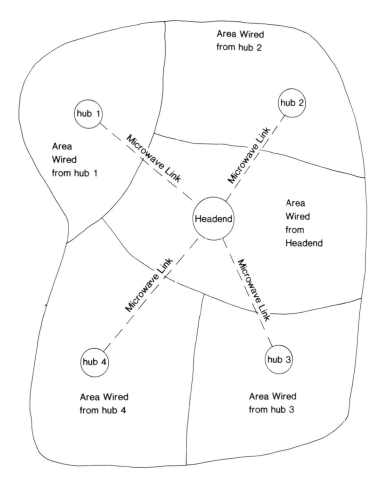

directly through the TV set. Subscribers are provided with converter boxes that enable them to receive the whole array of programming. Usually the viewers turn their TV sets to a particular channel and then use the converter, a small device with number on it, to change to the various signals supplied by the cable TV company.

Sometimes a subscriber does not want to pay for the whole array of programming, particularly if the cable TV company offers a number of different pay movie services. The channels that the viewer is not paying for must be blocked out. Sometimes this is done by placing an electronic device called a trap in the line to the viewer's home. This filters out the frequencies of the unwanted channel so it does not reach the TV set. In other instances channels can be activated or deactivated through interactive technology located at the main cable TV company office.[1]

The technical aspects of cable TV are complicated and costly. Many cable companies do not plan on breaking even until their seventh or eighth year. The amount of capital needed to build a cable system is enormous. Once it is built it is fairly inexpensive to operate unless technology has progressed so rapidly that the system is hopelessly out of date and must be rebuilt.

Although broadcasting suffers from the line-of-sight problem and has limited channels, it does not have the complicated distribution process of cable and does not have the degree of lead time until profit becomes a possibility.

## Cable Programming

The field of cable TV programming has been a fluctuating one. Early programming was mainly retransmission of already existing signals from TV stations. Occasionally early cable TV systems undertook their own production, but it was of a very limited nature.

Home Box Office, which began satellite transmission in 1975, was the first offering to become a national cable TV programming service. By the early 1980s, the number of national programming services had grown to approximately fifty depending on the moment at which they were counted. For a brief period of time, new services were announced almost daily by those wishing to cash in on the pot of gold in the sky. This proliferation of services was not economically sound, and soon even some of the most highly touted services folded or merged with other highly touted services.

The financial backing for cable programming has come from a wide variety of sources including traditional broadcast and entertainment companies, cable TV system operators and companies, oil companies, religious organizations, and some enterprising entrepreneurs. Many of the services have been co-financed by two or more entities, and some of these alliances have involved strange bedfellows who often line up on other issues as enemies. Some of the mergers and buyouts have been equally unorthodox with tiny companies sometimes challenging and defeating mighty giants in the tradition of the mouse swallowing the elephant.

Over the years, cable TV programming has evolved into four distinctive types: pay-cable, basic cable, local programming, and auxiliary services.

The main staple of pay-cable is movies. Subscribers pay a fee above and beyond the normal monthly cable TV fee in order to receive these services. HBO, Showtime, and The Movie Channel have been the three leaders in the pay-cable movie area, although smaller services such as Cinemax and Home Theater Network have survived for a number of years.

Another pay-cable service is the Disney Channel which features family entertainent, much of it coming from the Disney library. On the opposite end of the spectrum is the Playboy Channel which features "R" rated and unrated movies as well as interview shows and other programming aimed toward "adult" audiences.

Pay-cable services have also been tried or planned in the cultural and sports areas, but they have not had successful longevity. Similarly, attempts at pay-per-view (special events for which subscribers pay a one-time fee) have not been overly successful on cable TV.

The basic cable TV services are those for which consumers are charged very little above the regular monthly cable TV fee. Most of the services are attempting to be advertising supported and most select a particular type of programming to emphasize. The term narrowcasting, as opposed to broadcasting, has been used to describe this cable programming because the programming is intended for a narrow audience interested in a specific subject matter.

For example, ESPN features primarily sports; Cable News Network is twenty-four hours a day of news; Nickelodeon is for children; C-SPAN cablecasts live coverage of the House of Representatives and other public affairs type programming; MTV is twenty-four hours a day of rock music videos; The Learning Channel features educational material; ARTS covers the cultural scene; Lifetime consists primarily of health and lifestyle programs; Black Entertainment Television (BET) features black artists and issues; Spanish International Network (SIN) is totally Spanish programming; and quite a few services program totally religious material.

Another group of basic services are the super stations. These are regular broadcast stations that beam their signals to a satellite transponder as well as over a regular broadcast transmitter. The most famous of these is WTBS in Atlanta owned by Ted Turner. It was the first super station and was followed by WOR in New York and WGN in Chicago. All these stations are independents that program movies, local news, and sports.

Very few of the basic services are turning a profit. Their attempts at garnering advertising support have not been as successful as hoped. But, many of the concepts have strong backers with deep pockets who are willing to gamble on future profits.

Local programming is the third area of cable programming. This covers all programming created by a local cable system as opposed to that retransmitted from the satellite services. Some cable systems engage in local origination, some support access programming, and some do both. The difference between local origination and access is that l.o. programming is produced by people who work for the cable system, and its content is controlled entirely by the system. Access programming is produced by members of the community, usually in conjunction with the cable system, but the community members determine the content.

Local origination programming usually consists of news programs about the local community, local sports, and/or feature shows about local events. Some cable systems operate a local origination channel as though it were a local broadcast station. They sell ads and program both local and syndicated material to the extent that the income from ads allows. Different cable systems have different types of access channels. The most common are called public access and are to be used by any members of the community or non-profit organizations that wish to produce programs. Educational access channels are utilized by schools and colleges in the cable system's area, religious access channels are set aside for local religious groups, and government access channels have their use determined by the local governmental bodies. Leased access channels are for local groups that do not qualify as non-profit entities or individuals but who wish to see that a particular message is brought to the community. This would include local merchants or newspapers who may wish to simply present a message or may want to sponsor a particular type of program. Sometimes public access programs which are successful enough can obtain sponsors and become leased access programs. In this way the people involved can receive money for their labors.

How these channels are organized and operated differs widely from community to community. Sometimes the cable company will provide equipment, studio space, and professional personnel to help the various groups create their programming and will then be responsible for seeing that the programs are cablecast over the system. In other places the cable company will donate equipment to a community non-profit access organization, the city council, or a local school district. The organization will then produce its own programming with its own hired or volunteer staff. Other cable companies merely make a channel available for programming and the local organizations or individuals interested in doing the programming must find their own resources for equipment and crew.

Auxiliary services include a vast array of potentials as well as some services already in effect. Home security, video games, computer hobbyist interaction, and videotext are among the possibilities. Most of these auxiliary services involve taking advantage of the two-way nature of cable. Security systems can signal back to a central location when something is awry. Video games can be requested by the subscriber from a central computer and sent to the subscriber's terminal. People with computers can communicate with other computer owners through the cable system's send and receive wires. Through two-way videotext, subscribers can shop, bank, and carry on other daily transactions.

This vast array of cable programming has made a dent in the viewership of the networks. However, people with cable TV apparently watch more television overall than people who have not subscribed to cable. All of the programming is so new that its future course is hard to predict, but at present there is room for experimentation and innovation. The programming concepts of cable are still fairly new. Some have been quite successful, and others may still see their hey day.[2]

**Financial Considerations**

Unlike broadcasting, cable TV systems must sell their services to the consumer. This means that someone from the cable TV organization must contact individuals and convince them they should subscribe to cable TV. In many cases this is accomplished through direct old-fashioned door to door selling. Because many people are not home during the day and because people are increasingly leery about letting strangers in their home, the cable companies often hang notes on people's doors first telling them that a sales representative will be calling in about a week.

Because of all the publicity given to cable TV, some people are eager to buy and have an understanding of what they are buying. But many others are ignorant of cable TV in general or of the various options that cable provides. This means that sales people must educate as well as sell.

A primary difficulty with selling cable subscriptions, especially on the newer systems, is that there are many different pricing structures depending on which programming channels the viewer wishes to receive. To handle this, cable systems have developed a system called tiering. Although different systems tier in different ways, most offer a bottom tier that consists of local stations and local access channels for a minimal price in the neighborhood of three to ten dollars. For a few dollars more, the consumer can have all of the bottom tier 1 plus programming included in tier 2, perhaps the super stations and two other basic cable channels. Tier 3 would consist of all of tiers 1 and 2 plus an additional 15 or so of the basic channels. This, too, would cost only a few dollars more. Tier 4 would include all of the local and basic programming plus one pay-TV service which usually costs an additional ten to twenty dollars. Tier 5 would include two pay services and Tier 6 would include three. With each of these additional pay channels the price would rise significantly. The customers choose the level of service they wish, pay a one-time installation fee, and then pay on a monthly basis for the services for which they contracted.

A few cable systems are also experimenting with pay-per-view whereby subscribers pay to watch one particular program or special sports event. This further complicates the payment structure but may become more common in the future as equipment that enables cable companies to switch signals off and on in each subscriber's home becomes more refined.

Most of the income circulating through the entire cable TV industry comes from payments from subscribers. And this can be a sizeable amount. For example, a cable system with 100,000 subscribers each paying an average of $20 a month generates an income of $2 million a month—a definitely respectable amount. However, advertising also provides some funds. Most of the basic programming services are soliciting advertising and some of them, such as ESPN and CNN, have been quite successful in their endeavors.[3]

At present, the cable systems pay the cable network program suppliers for the programming. The amount is usually several cents per subscriber per month for the basic services and several dollars per subscriber per month for the pay-cable services. One of the main ways cable systems decide which of the multitude of services they want to place on their channels is to consider the popularity they feel the channel will have in their community in relation to the amount they must pay the cable network.

Some basic services, particularly the religious ones, are offered to cable TV systems free. The Spanish International Network pays cable systems a set amount for each Spanish surname on their subscriber list, leading it to develop the slogan, "SIN Pays." Other cable network programmers have indicated that they, too, may pay the cable systems when they attract enough advertising to cover their expenses.

Attracting advertisers for cable TV has become a similar game to attracting advertisers for broadcast TV. Advertisers are interested in numbers, both the total number of households that are available to view their ads plus the number who actually do. The total number available relates to a figure the cable industry refers to as penetration. This is the percent of homes in the U.S. that subscribe to cable TV. When cable penetration reached 30 percent, advertisers became much more interested in cable.

Penetration varies from area to area. Outlying areas with poor broadcast reception or only a few channels have proven to have higher subscription percentages than large, flat cities with a multitude of TV stations available.

Because of the large number of channels available on many cable systems, the audience is fragmented, and the ratings methodology to determine the actual number of homes viewing the smaller services has not been refined. However, because of cable's narrowcasting approach to programming, advertisers can reach audiences more efficiently. They know that people watching a particular channel have particular interests and can select to advertise on channels that have programming that dovetails best with their products.

Local origination channels are also aiming for advertising dollars from local merchants. Some have been successful by offering low rates, which are in line with the rates charged by local radio stations or newspapers.

The overall financial state of cable is a bit uncertain. Not many of the companies are making large profits. Many have dropped by the wayside, and others have retrenched and consolidated. Cable TV is not a dying business, but it has not lived up to its financial predictions.

### Regulation

Regulation of cable comes from a variety of government bodies that have, from time to time, shown a reluctance to regulate cable.

The Federal Communications Commission regulates signals which travel through the air. Because cable TV is primarily on wire, the FCC does not have clear-cut jurisdiction, and much of what it has done in the cable TV regulation area has been the result of backing in through broadcast regulation. In other words, in its attempts to tighten or loosen the reins of broadcast TV, the FCC has made rulings regarding such issues as the rebroadcasting of TV station signals which, in turn, affect the cable industry.

Congress has passed numerous bills which affect cable TV, but it is not a constant guardian of the industry. It has dealt in significant ways with copyright provisions regarding cable TV and has passed a bill which dealt specifically with the regulation of cable TV by local governments.

The local governments have been the primary regulators of cable TV because they have been involved in awarding and renewing franchises. Cable TV companies must receive franchises before they can begin wiring in a particular area and must renew these franchises periodically in order to stay in business. Because the wiring is local in nature, it is usually the municipal government which gives this permission, although in some places the state is the governing body involved.

When a local government decides it wants to have cable TV in its area, it issues a "request for proposal" stating that anyone interested may bid to become the cable franchise holder. Usually within this RFP are certain stipulations that the city officials have decided they would like from the winning bidder. These may deal with the number of channels, the areas where cable is to be laid underground as opposed to on telephone poles, equipment and staff that the cable company should provide for local programming, the length of time the cable company should spend building the system, and the percentage of its income the cable company will pay the city as a franchising fee.

Paying a franchising fee became highly controversial in the early 1980s because cable systems, eager to receive franchises, were considering promising cities unrealistically high amounts that, in essence, gave away the store and made it very difficult for cable companies ever to breakeven. The FCC stepped in and ruled that no cable system could offer a city more than 5% of its income.

Any company that wishes to bid for the franchise obtains a copy of the RFP and writes up a proposal stating what it will do for the city if granted the franchise. In order to write this proposal many of the companies have had their employees interview people in the community to determine how the city's needs can best be met. Others have placed community leaders on boards to advise the cable company regarding cable activities. Still others have given shares of stock to local leaders. This latter practice, dubbed "rent-a-citizen" came under sharp criticism and has been largely discontinued; in general, companies have done whatever they felt they needed to, both formally and informally, in order to win the franchise.[4]

The RFPs include a due date by which companies must file for the franchise. On that day or before, the companies bring the written proposal, usually several volumes thick, to the city hall. City employees or consultants then review all documents submitted, summarize data, and make preliminary recommendations regarding which candidate should win. At this point the cable companies are usually given a chance to reply to any comments made on their proposals, and sometimes they are even allowed to make changes in what they proposed.

The main city governing body, usually the city council, is then charged with making the final decision. The council may decide to hold a public hearing to which interested citizens can come and give their input, or it may simply review the summarized data. Usually, the council agrees with the consultant or staff recommendations, but such is not always the case. Sometimes the cases become quite muddy, particularly if different companies have been favored along the way, or if there are cries of scandal or corruption.

In some instances franchising battles have wound up in the court as disgruntled losers cry unfair favoritism on the part of the winners.

Once a company has been granted a franchise, it has a virtual monopoly on the cable TV service within that particular community. Many cable franchises are not specifically non-exclusive, and theoretically a second company could apply to wire the same area as the company that won the franchise, but such is not a common practice. The cost of wiring and operating a cable system is such that the likelihood of two companies surviving economically is not very great.

The winning company is given the franchise for a set number of years, usually fifteen. At that time it must apply to the city for franchise renewal. If it has not lived up to its agreements with the city, then its franchise can be withdrawn and given to another company. This has happened rarely.

More common is the taking away of a franchise long before the franchise period is up. This happens when companies win the franchise but do not build the system in the time allotted by the city. Cities try to convince the franchise holder to build more rapidly, but if the situation becomes stagnant, the city has the right to reopen the bidding process and choose another company.

Regulation of cable has been both controversial and confusing. Recent regulation acts have been more organized than early ones, but the field is still in flux.

## History

Cable TV has been in existence almost as long as broadcast TV, but it waited a long time to become a household word. It experienced modest beginnings, slow progress, an unprecedented boom, and a return to reality.

### The Very Early Days of Cable

There are many different stories about how cable TV began. One is that it was started by a man in a little appliance shop in Pennsylvania, who was selling television sets. He noticed that he was selling sets only to people who lived on one side of town. Upon investigation he found that the people on the other side of town could not obtain adequate reception, so he placed an antenna at the top of a hill, intercepted TV signals, and ran them through a cable down the hill to the side of town with poor reception. When people on that side of town would buy a TV set from him, he would hook their home to the cable.

Another story is that cable service was started by a "ham" radio operator in Oregon who was experimenting with TV just because of his interest in the field. He placed an antenna on an eight-story building and ran cables from there to people's homes. The initial cable subscribers helped pay for the cost of the equipment, and after that was paid off, they charged newcomers $100 for a hook-up.[5]

Whether or not these stories represent the true beginnings of cable TV is hard to say. But there were many remote or mountainous places where friends and neighbors gathered together with the intent of providing television reception for themselves.

One factor that helped cable TV in its beginning was the freeze that the FCC imposed on broadcast television stations from 1948 to 1952. During this time, while the FCC was deciding how the station frequencies should be allocated, no new TV stations were allowed to be started. The only way for people to receive TV if they were not within the broadcast path of one of the 108 stations already on the air was to put up an antenna and, in essence, catch the signals as they were flying through the air.

Most of the early cable TV systems were capable of handling three signals. As early as 1949 a multi-channeled antenna system was developed in Lansford, Pennsylvania as a money-making venture. With no local signal available, the system carried the three network signals by importing them from nearby communities, a practice which became known as distant signal importation. Fourteen such signal importation companies were in operation by the end of 1950, and the number swelled to seventy by 1952.[6]

At this point the FCC was convinced that as the number of stations increased, the need for cable TV would diminish and gradually vanish entirely. The one factor that eluded their attention was that many communities were too small to support the expense of station operation. If the basic philosophy of "mass communication for an informed public" was to be a reality, cable would have to grow. Signals would have to be imported from cities where stations could find support for their operations.

However, even with 65,000 subscribers and an annual revenue of half a million dollars in 1953, cable TV was still only a minor operation. Most broadcasters felt no concern about this business, which was growing on the fringes of their signal contour. Some, however, were becoming alarmed by the feeling of permanence growing in some cable systems. Coaxial cable was replacing the open line wire of early days; space was leased on telephone company poles for line distribution instead of the house-to-house loops augmented by a tree here and there.

Placing the cable on phone poles led to one of the major thorns in the side of the cable industry. The cable companies had to rent space on the phone poles for their cable, and because the phone companies naturally had a monopoly on all the telephone poles in town, they could charge rates that the cable companies felt were unfair. The cable companies also felt that the phone companies often gave them an unnecessarily hard time concerning where, how, and when the cables were to be attached on the poles. Much discussion and lobbying revolved around pole attachments, but the issue has not yet been satisfactorily resolved.

As cable TV grew, it became a little more sophisticated and, in addition to importing signals into areas where there was no television, began distant signal importation into areas where there was a little television. For example, a small town might have one TV station, and the cable system would import the signals from two TV stations in a large city several hundred miles away.

In general, during the decade of the 1950s cable TV grew from a few very small operations of friends and neighbors to a system of television reception for small and medium-sized areas, which had poor reception or a very limited number of TV stations. Most of these systems were operated by small locally run organizations that charged subscribers an initial installation fee and monthly fees compatible with modest profits.[7]

## Confused Growth

It was the importing of distant signals that brought about the first objections to cable TV. The existing TV stations in the area would find that their audience had shrunk because people were watching the imported signal. In fact, sometimes the signal imported would be the same as the one on the local station. For example, the local station might be showing a rerun of "I Love Lucy" and find that the imported station was showing the same rerun, splitting the audience for the show in half. The result of all this was that local stations could no longer sell their ads for as high a price as they had before the importation.

Some of the stations in areas affected by the cable TV appealed to the Congress and the FCC to help them in their plight. In 1959 the Senate Commerce Subcommittee suggested legislation to license cable operators. The actual extent of cable operations, even at this late date, was impossible to identify because the operators were not required to report to any governmental agency—in spite of the fact that in many areas of the country the audience served by cable ran as high as 20 percent of the available viewers. The attempts to draft legislation were littered with arguments and debates lasting into 1960 and ending with the defeat of a bill proposed by Senator John Pastore.[8]

With the failure of federal intervention came a rash of state and local attempts to assert jurisdiction over cable TV. In most areas the local city council became the agency that issued cable franchises, largely through default of other government bodies.

During the '60s franchising was a very quiet process usually with only one or two companies applying for a franchise in a given area. Areas with good reception did not bother with franchising and probably could not have even given away their franchises if they had wanted to.

Although the number of cable systems doubled between 1961 and 1965, cable TV was still a small business. In 1964 the average system served only 850 viewers and earned less than $100,000 annually. Multiple system ownership, the owning of a number of cable TV systems in different locations by one company, was less than 25 percent because of the lack of economic incentives.[9]

This lethargy was not the case in some foreign countries where cable caught on more quickly than in the United States. In Canada, for example, the political and geographic climates were such that most of Canada was wired before the U.S. even considered cable an important entity.

In America, however, the FCC maintained a policy of nonintervention in cable TV matters during the early '60s. It was hoping that the problems between the operators and broadcasters would be settled by court decision. Unfortunately, the situation only became more confused as court cases piled upon court cases.

The early '60s, then, saw the beginnings of a rift between broadcasters and cable TV people over distant signal importation, a lack of action on the part of the federal government, and the beginning of actions on the part of local governments.

## FCC Actions

In April of 1965 the FCC did act and issued a "Notice of Inquiry and Notice of Proposed Rulemaking." In its "First Report and Order," it covered only two main areas. (1) All cable linked common carriers from this time forward would be required to carry the signal of any TV station within approximately sixty miles of its system. (2) No duplication of program material from more distant signals would be permitted fifteen days before or fifteen days after such local broadcast.

The rule of local carriage, which became known as the must-carry provision, caused little or no problem. Most cable systems were glad to carry signals of local stations. But the thirty-day provision caused bitter protest from cable operators for it limited their rights in relation to what they could show on their distant imported stations. This rule, which became known as syndicated exclusivity, meant that a cable TV company could not carry a syndicated show on a distant station if a local station had a contract to broadcast the show sometime within fifteen days before or after the distant station was scheduled to broadcast the show. So, if a local station was going to show the "I Love Lucy" rerun on January 15, a cable TV operator would have to black-out that rerun on a distant imported station if it were on any time between the beginning and end of January.

The cable TV industry marshalled its forces and in May of 1966 succeeded in having the thirty-day provision reduced to only one day. This, of course, angered the broadcasters.

As background to this conflict, one must visualize the frustration of broadcasters who had spent large sums of money. Many were barely able to survive on the revenue generated by their station while cable operators with far less invested were realizing profits by carrying the broadcasters' signals and importing distant signals. During this period there were on file and pending applications for cable coverage of areas both urban and rural, which would account for at least 85,000,000 people. It is easy to understand the fears of station operators. Their basic desire was for the FCC to provide them with full administrative protection.

The attitude at the FCC during this time was strongly in favor of local TV stations in every area of the country. The second report issued by the FCC in 1966 restricted cable service in the top 100 markets to existing services. This order came when 119 systems were under construction, 500 had been awarded franchises, and 1200 had applications pending. All of these systems would be required to prove that their existence would not harm any existing or proposed station in their coverage area. By making no increase in staff to handle this load, the FCC was, in essence, freezing the growth of cable operations in the top 100 markets. This made station owners very happy.

The effect this ruling had on cable operators during the remaining years of the decade was the reverse of what the FCC had in mind. Long on advocate of local service and ownership, the FCC could not or did not foresee what would happen. Cable systems, unable to expand, were sold in large numbers to large corporations, which could withstand the unprofitability of the freeze period. Multiple station ownership became quite prevalent by 1970.

In 1972 the FCC issued another policy on cable TV, which changed the "must-carry" rule slightly from all stations within sixty miles of the cable TV to all stations within thirty-five miles plus other local stations, which, as shown by polls, were viewed frequently by people in the area. The policy also reconfirmed syndicated exclusivity and set up guidelines delineating how much distant signal importation a cable system could undertake depending primarily upon the size of the market where it was located and the number of stations already in the market. The FCC also established four different kinds of channels that should be on a cable system:

The class 1 channels were retransmission channels, which served the function cable TV had been involved with since its inception—merely providing programming already on the air.

Class 2 channels were the local origination channels, which cable systems were to use to originate programs.

Class 3 channels were coded channels, which could not be received unless the subscriber had special equipment installed to unscramble the signal. This was intended for special programs that the viewer would have to pay extra for in order to watch.

Class 4 channels were to provide for two-way communication so that a messge could be sent from the subscriber's home to the cable origination point or to some other destination.[10]

Not much was accomplished regarding these channel requirements, and for many years they were treated with benign neglect. In fact, during the remainder of the 1970s the FCC withdrew further and further from cable regulation leaving it more and more in the hands of local municipalities.

**Early Programming**

When cable TV first began it was a common carrier similar to the telephone company; i.e., it picked up signals and brought them into homes for a hookup fee and a regular monthly fee. This meant that the programming was only that which was picked up from local stations or from distant signal importation. Therefore, the early programming on cable TV was programming of ABC, CBS, and NBC, as well as various local broadcast stations.

Under this early system there was no local origination of programs. But gradually some of the cable facilities began undertaking their own programming. The most common "programming" in the beginning was unsophisticated weather information. The cable TV operators would place a thermometer, barometer, and other calculating devices on a disc and have a TV camera take a picture of it as it rotated slowly. This would then be broadcast on one of the vacant channels so that people in the area could check local weather conditions. Some systems had news of sorts. This might just be a camera focused on a bulletin coming over a wire service machine, or it might be three by five cards with local news items typed on them. At any rate, it was simple, inexpensive, one-camera-type local origination.

Gradually, studio-type local origination began usually in the form of local news programs, high school sports events, city council meetings, local concerts, and talk shows on issues important to the community. The first regularly scheduled cable local origination was in 1967 in Reading, Pennsylvania, and shortly after that there was local programming in San Diego, California. The FCC became involved in this and gave the San Diego cable system the authorization to engage in local programming. After that other cable systems began such programming.[11]

In October of 1969 the FCC issued a rule that required all cable TV systems with 3500 or more subscribers to begin local origination no later than April of 1970. The purpose was to promote local programming for the public in areas where it did not exist.

When April of 1970 came along many of the cable TV operators who were not engaging in local origination claimed hardship and told the FCC that they did not have the funds to build studios, buy equipment, and hire crews. As a result, the FCC's order was not enforced. In 1971 the FCC made another attempt to require cable systems to engage in local origination. This too failed, so the FCC decided only to require that systems with over 3500 subscribers make equipment and channel time available to those wishing to produce programs.

This brought about a different type of local programming, which became known as public access. Individuals or groups within the community would use the equipment provided by the cable operator to produce programs and then would cablecast those programs over one of the system's channels.

In most areas public access was not a success and the equipment provided by the cable company sat on the shelf for lack of interest on the part of the local community. In other areas, particularly where cable systems had shown an active interest in local programming, some exciting and innovative projects were undertaken in the public access area that included public affairs programs and video art. Unfortunately, in several places the people who made use of the public access time were would-be stars using cable for vanity purposes or people from fringe groups wishing to propagate various pressure group causes or even lewd modes of behavior. These individuals or groups made it difficult for the cable operators to present balanced opinions and also gave the concept of public access an unsavory reputation.[12]

Another form of local programming instituted by some cable systems was the showing of movies. The cable company would use one of its channels or perhaps only the evening hours of one of its channels to show movies, which only subscribers who paid a fee above and beyond the regular monthly fee could receive.

These movies were leased from film companies and were shown without any commercial interruptions. There were regulations to keep cable systems from showing the well-known films that broadcasters wanted to show during prime time, but because the small cable systems only had a small number of subscribers who would pay for the movies, they could not afford to pay for blockbuster movies anyhow. The quality and vintage of most cable movies was similar to what could be seen on late, late broadcast TV. The channels for movies were not a huge success, but they did bring in extra income to some cable systems.

By the mid-70s only 20 percent of the cable companies were engaging in any type of local programming, and the most popular type of such programming was local sports.[13]

However, throughout the '60s and '70s, promises were made, broken, and re-made concerning the potential services and programs that would be available through cable. Cable lived on the edge of a promise that "within the next five years" cable TV will perk your coffee, help your kids with their homework, balance your checkbook, bring you the top sports events, secure your home, teach the handicapped, and do your shopping. But, except for a few isolated experiments, cable remained, until the mid-70s, primarily a medium to bring broadcast signals to areas with poor reception.

## Copyright Controversies

Another sore point between broadcasters and cable operators surrounded the payment of copyright fees. When cable systems rebroadcast network, local station, or distant station signals, they did not pay any fees to those who owned the copyrights to the materials. In some instances the networks or stations had paid for and created the material so owned the copyright. In other cases the networks or stations had purchased or rented program material from film companies or independent producers and had paid copyright fees as part of the overall payment.

The stations, networks, film companies, and independent producers felt cable companies should pay copyright fees for the retransmission of material because these retransmission rights were not included in the broadcast TV package. Cable operators felt that copyright had already been taken care of by the TV stations and networks, and cable systems were merely extending coverage.

The basic fight over copyright began in 1960 when United Artists Productions, holders of license rights to a large library of feature films purchased by a network at a cost of $20,000,000, sued two cable operators for copyright infringement of their performance rights. United Artists cited the 1931 case of a hotel owner who was required to pay royalties for radio music programs piped to rooms of his hotel. The initial ruling handed down in 1966 favored the plaintiff, United Artists. On appeal to the U.S. Supreme Court, the decision was overruled in 1968, and the controversy once again boiled over.

Further complicating the issue was the fact that the copyright law under which the United States then operated was written in 1909. Congress had been trying for many years to update the copyright situation with new legislation, but the many controversies and special interests surrounding the issue, including the cable feud, made this one of the most lobbied bills in history.

Finally, a new copyright law was passed in 1976 and went into effect in 1978. Under this law cable TV systems were to pay a compulsory license fee to a five-member newly created government body called the Copyright Royalty Tribunal which would then distribute the money to copyright owners. The amount of this compulsory license fee was 0.7% of a cable operator's revenue from basic monthly subscriptions. The copyright law gave the Tribunal authority to adjust the rate for inflation, and cable operators and copyright holders seemed happy with this plan.[14]

Two major problems arose. One problem involved the distribution of the money collected through the compulsory license fee. Interests whose program material was retransmitted over cable claimed a share of the money, and it was the Tribunal's job to decide who got how much. In 1979, the Tribunal decided that the $20.6 million collected should be distributed as follows: 70% to movie producers and program syndicators; 15% to sports; 5.25% to public television; 4.5% to broadcasters; and 0.25% to National Public Radio. Hardly any of the entities was satisfied with its piece of the pie and most went to court in an attempt to gain increases.[15]

The other problem arose because the cable industry changed greatly after 1976 and the $20.6 million collected through compulsory licensing seemed like a paltry sum to copyright holders. Cable systems with expanded channel capacity were retransmitting signals from stations all over the country but, according to copyright holders, were paying as though they were only retransmitting the local must-carries and a few distant imports.[16]

By the early 1980s there was seething discontent with the infant copyright law. Lobbyists from many entities were trying to convince Congress to initiate bills to better their lot in life.

## The Cable-Satellite Alliance
In 1975 the stage was set for a dramatic change in cable programming when a company called Home Box Office began distributing movies and special events via satellite.

Home Box Office was actually formed in 1972 by Time Inc. as a pay movie/special service for Time's cable system in New York. The company decided to expand this service to other cable systems so set up a traditional broadcast-style microwave link to a cable system in Wilkes-Barre, Pennsylvania. In November of 1972 HBO sent its first programming from New York to Wilkes-Barre—a National Hockey League game from Madison Square Garden followed by a film "Sometimes A Great Notion."

During the next several years HBO expanded its microwave system to include about fourteen cable companies in two states with over 8000 subscribers. This was not an overly successful venture and was not profitable for Time Inc.

Then, in 1975 shortly after domestic satellites were launched, HBO decided to bring two of its cable systems the Ali/Frazier heavyweight championship fight from Manila by using satellite transmission. This experiment was very successful and HBO decided to distribute all of its programming by satellite.[17]

This was easier and cheaper than the system that required huge microwave towers, and it meant that as soon as HBO sent its signals to satellite, they could be received by any cable system in the entire country that was willing to buy an earth station satellite receiving dish.

HBO then began marketing its service to cable systems nationwide, no easy chore at first. The original receiving dishes were ten meters in diameter and cost close to $150,000, a stiff price for cable systems, many of which were just managing to break even. But the technology of satellites moved quickly enough that by 1977 dishes that could be as small as 4.5 meters sold for under $10,000.[18]

Another problem HBO encountered was that rules had been established, mainly for over-the-air subscription TV, that prohibited pay services from bidding on movies and sports events that conventional broadcasters wanted to show. These rules had been established to protect broadcast TV stations from a phenomenon known as siphoning. The fear was that pay TV might simply take over, or siphon, the TV station's programming by paying a slightly higher price for the right to cablecast the material. HBO and several cable TV system owners took these rules to court, and in March of 1977 the court set aside the pay cable rules restricting programming, leaving HBO free to develop as it wanted.[19]

When HBO first began marketing its service, it offered the cable system owners 10 percent of the amount they collected by charging subscribers extra for HBO programming. Approximately 40 percent of the fee was to go to HBO and 50 percent to the program producer. When cable owners indicated they were not ecstatic about their percentage, HBO raised it so that the systems retained about 60 percent of the money with HBO and the program producers splitting the other 40 percent.[20]

With receiving dishes manageable both in terms of cost and size, with programming of an appealing nature, and with financial remuneration at a high level, cable systems began subscribing to HBO in droves. Likewise, HBO became very popular with individual cable subscribers, and by October of 1977, Time Inc., was able to announce that HBO had turned its first profit.

Shortly after HBO beamed onto the satellite, Ted Turner who owned a low-rated UHF station in Atlanta, Georgia, decided to put his station's signal on the same satellite as HBO so that cable operators who had bought a receiving dish to receive HBO would also be able to cablecast his station. This station became known as a super station because it could be seen nationwide. The carrier of the station charged cable operators a dime a month per subscriber for the signal[21] but they, in turn, did not charge the subscriber as they did for the HBO pay service. The economic rationale for the super station was that the extra program service would entice more subscribers. On the other side of the coin, the station was able to charge a higher rate for its advertisements now that it had a bigger audience spread out over the entire country.

With two successful program services on the satellite, the flood gates opened, and cable TV took on an entirely new complexion.

## Cable's Gold Rush

In the late 1970s and early 1980s cable TV experienced a phenomenal growth, sparked mainly by the development of satellite delivered services. In 1975 265,000 homes subscribed to pay-cable, but that number grew to one million by 1977, over six million by 1980, and eight million by 1981.[22] Similarly, revenues from pay-cable increased. In 1979 pay revenues grew 85 percent over 1978, and the industry predicted that was a peak, which would subside to about 50 percent growth.[23] But in 1980 pay revenues grew 95.5 percent over 1979.[24] The number of subscribers to cable, in general, also grew. During the 1970s cable added subscribers at the rate of about 1.1 million per year, but in 1980 alone 3.1 million subscribers were added.[25] Between 1975 and 1980, the percentage of homes subscribing to cable jumped from 17 to 23 percent,[26] and during this same period of time the profits of cable TV grew 641 percent.[27] With figures like that, it is no wonder that cable experienced a veritable gold rush.

One of the main areas where this gold rush manifest itself was in the area of franchising. The cities that couldn't have given away their franchises in earlier years because their reception of TV signals was so clear suddenly became prime targets for cable and its added programming services. In general, cabling moved from rural areas into major cities such as Pittsburgh, Boston, Philadelphia, Los Angeles, Dallas, and Cincinnati.

A minor franchising boom had occurred in urban areas in the early 1970s when predictions abounded that cable would be growing rapidly, but little had come of that. Now, in the late 1970s, the franchising wars swung into full force with close to a dozen different companies applying for the privilege to wire large and small sections of urban and suburban communities.

The local governments charged with conducting cable franchising were generally not accustomed to dealing with matters of this fast-paced communication nature. Gradually, through the use of consultants, they established lists of minimal requirements they wanted from the cable companies and developed processes for selecting franchise winners.[28]

Cable companies, in their fervor to obtain franchises, usually went well beyond what the cities required. They, too, hired consultants who contacted city leaders to learn the political structure and needs of the city and to decide how the company should write its franchise proposal to insure the best possible chance of winning. Cries of scandal sometimes accompanied these franchising procedures, as the cable companies tried to gain influence.[29]

As franchising competition became more intense, cable companies began promising cheaper rates, shorter time to lay the cable, more channels, more equipment, more local involvement, and generally, more and better everything. Sometimes the winning cable company was unable to meet all the requirements stipulated in its bid, especially in regards to the speed with which the system was to be built. This led to fines and court cases over breech of contract.[30]

But cabling did move forward at a rapid rate, and as the cable companies began promising more, they realized they might not recover their investment for about a decade. This hastened a process already in full swing in the cable industry—the take over of small "ma and pa" cable operations by large multiple system owners (e.g., Teleprompter, ATC, Warner, TCI, Cox, Storer, Times Mirror, Rogers)[31] and then the consolidation of these MSOs with other large companies, (e.g, Teleprompter with Westinghouse, ACT with Time Inc., Warner with American Express, Rogers with UA-Columbia, Cablecom General with Capcities).[32] Large companies emerged in cable partly because they wanted to be part of the gold rush and partly because only large companies had the resources to withstand the expenses of the franchising process and the other start up costs of laying cable, marketing, and programming.

Other groups also began to stake their claims in cable's gold rush. Advertisers who saw cable reaching close to 30 percent penetration of the nation's households, became interested and began placing ads on cable's programming channels, both local and national. Sports programming was the first to attract advertising, but soon corporations were committing to advertising for other areas of programming such as news, health, culture, and entertainment.[33]

Members of the various unions and guilds that operate in the broadcasting industry were not involved in cable programming when it first began because the cable companies did not recognize the unions or abide by their regulations. The early original cable programming was written, directed, produced, and crewed by non-union members. However, after several long, bitter strikes in the early 80s, the unions won the right to be recognized and paid, and the same people who worked in broadcast programming began working in cable programming.[34]

Perhaps the biggest winners in the cable gold rush were the equipment manufacturers who supplied the materials needed to build the cable systems. The suppliers of the converters, which enable a regular TV set to receive the multitude of cable channels, the earth station dishes, and the cable itself found their order desks piled high.[35] Space on a satellite became a precious, premium commodity as more and more companies wanted to launch national programming services. Satellite transponder time, which had rented for about $200 an hour or $1 million a year,[36] took on a new dimension when RCA auctioned off seven transponders on its Satcom IV satellite for figures ranging from $11.2 to 14.4 million for six years. The FCC later disallowed the auction method of leasing transponders, and RCA had to settle for charging all successful bidders $13 million for seven years and nine months of use.[37]

Although large companies, advertisers, unions, and equipment manufacturers all became involved in cable during the late 1970s and early 1980s, the most generally noticeable type of growth was in the area of programming and interactive services.

## Development of Pay Cable

Shortly after Time Inc. succeeded with its HBO service, Viacom launched a competing pay-TV service on satellite called Showtime. Viacom, like Time, owned various cable systems throughout the country and had been feeding them movies and special event programming through a network that involved bicycling and microwave. It joined with another cable system owner, Teleprompter, to own and launch Showtime.[38] Teleprompter was subsequently purchased by Westinghouse, which divested itself of the Teleprompter interest in Showtime by selling it to Viacom, making it the sole owner.

Following the launch of Showtime, Warner Communications in conjunction with American Express company (Warner-Amex) introduced the Movie Channel which consisted of twenty-four hours a day of movies.

Time Inc., buoyed by its success with Home Box Office, began another pay service called Cinemax consisting mostly of movies programmed at times complementary to HBO.

Galavision went on the satellite as a Spanish-language movie service, Times Mirror established Spotlight as a movie and special service to its cable companies, and several companies entered the field of cultural pay-programming in the form of Bravo (owned by Rainbow) and RCTV (owned by Rockefeller Center and RCA).[39] Disney began a family-oriented service called The Disney Channel which offered old Disney films and newly developed materials.[40]

Several companies also entered the area of satellite delivered pay "adult" programming including R-rated movies, skits, and specials. Both Playboy and Penthouse entered this field, Playboy by joining forces with an already established program service, Escapade.

In what will probably remain as only an interesting footnote to the history of pay-cable, Getty Oil Company and four film production companies joined together in April of 1980 to form Premiere, a company which was intended to be a pay-TV movie service competing with HBO, Showtime, Movie Channel and Cinemax. The difference was that the Getty venture, in conjunction with Columbia Pictures Industries, MCA Inc., Paramount Pictures Corporation, and 20th Century-Fox Film Corporation, was going to show the films from those companies and then not allow them on any other pay-TV service for nine months. In essence, Premiere would have had half the top motion pictures nine months before any of the other pay services could exhibit them. Needless to say the other pay services fought this in the courts. After about a year of battling, Premiere closed its doors, the victim of two negative court decisions.[41]

## Basic Cable Growth

Many national satellite-delivered services developed during the late 1970s and early 1980s are neither pay TV services nor local stations that are merely retransmitted. Some of these are supported by advertising, some are supported by the institutions that program them, and some are supported by small amounts of money that the cable companies pay to the programmers. These services have been lumped together and referred to as basic cable because the subscriber does not usually pay any substantial additional fee for them.

Many of them are single purpose, programming one specific type of material. One of the earliest of these was ESPN which was originally owned by Getty Oil. It programed twenty-four hours a day of advertising based sports programming, featuring virtually every conceivable type of sport.

Another early service was Nickelodeon, owned by Warner-Amex, and founded on commercial free children's programming, some of it produced at cable systems owned by Warner Amex. Cable systems paid about 15¢ a subscriber a month in order to receive the service.

Another Warner-Amex entry was MTV which programmed music videos and news and information about the musical scene. Primarily country-western music was featured on Westinghouse's Nashville Network. This channel also included game shows and dramas in the country format.

Two cultural programming services were also offered, each owned by one of the major broadcast networks. ARTS, owned by ABC, was advertising based and received much of its programming from Europe. CBS Cable, also advertising supported, was owned and operated by CBS and featured a great deal of originally produced American material.

Competing services also appeared in news. The first, Cable News Network, was owned by the same Ted Turner who put his Atlanta UHF station on satellite. CNN programed 24 hours a day of news. It charged the cable systems to cablecast it and also solicited advertising. Satellite News Channel (SNC) was a Westinghouse-ABC joint venture of 24 hour news designed to compete with CNN.

CBN, the Christian Broadcasting Network headquartered in Virginia, provided free religious programming with the cost underwritten by donations from those who wished to contribute to their religion. Trinity Broadcasting Network (TBN) operated in a similar manner to other religious services such as the National Christian Network, the Jewish Network, and the Episcopal Television Network.

Satellite Program Network, based in Oklahoma was an unusual concept in that it served as a distribution vehicle for independently produced programs, some of which had their roots in public access. It charged the producers who, in turn, secured their own advertising to cover their costs. The programming was quite varied including interviews, sports, gardening, and women's programs.

Another unusual concept which became fairly popular was that designed by C-SPAN, the Cable Satellite Public Affairs Network. Based in Washington, D.C., this non-profit organization delivered live coverage of the House of Representatives for which it charged cable systems a modest fee per subscriber per month.

ACSN, the Appalachian Community Service Network, was originally underwritten by a grant to bring education to the Appalachian area. It later expanded to provide college credit courses for which it charged both the cable companies and the universities that gave credit for the courses.

SIN, the Spanish International Network was supported by advertising and provided Spanish-language programs. It paid the cable system for each Spanish surname on its subscriber list.

Cable Health Network was first owned by Viacom and was devoted to information about physical and mental health.

USA Network provided programming broader in scope than some of the other services. Although it began with a heavy emphasis on sports, it also included children's shows, health programs, and movies.

WTBS, WGN, and WOR were all "super stations" which carried commercials and also charged cable operators about ten cents a subscriber a month. WTBS was the Atlanta-based original super station; WGN was from Chicago; and WOR was from New York.

On a different track, Disco Network was an audio service fed over satellite. Although rarely discussed, cable is also capable of delivering radio stations and audio material to the home. Several other audio services were Home Music Store, WFTM Chicago, and National Classical Network.

And on and on. There were also services that provided words and numbers on the screen. For example, Associated Press, United Press International, and Reuters News Service had cable news services that printed out the latest news. Several companies offered stock market and other financial reports and some made weather information available twenty-four hours a day complete with drawings of weather patterns.[42]

For several years new pay and basic services were announced at a rapid rate—sometimes several in one day. Some of these never got off the ground; others existed only for short periods; and others showed signs of longevity.

## Changes in Local Programming

Because the franchising process placed so much emphasis on the local community in which the cable TV system was located, local programming took on an entirely new dimension in the late 1970s. The older systems that still had only a twelve-channel capacity usually allocated only one channel to local programming. The newly franchised systems able to take advantage of improved technology to provide twenty, then thirty seven, then fifty four, then over a hundred channels usually promised an entire complement of local channels.

At least one of these channels was generally reserved for local origination, programming the cable system itself initiated. The others were some combination of access channels—public access for the citizenry at large; community access for community groups such as the Girl Scouts and the United Way; government access for local officials; educational access for schools and colleges; religious access for religious groups; and leased access for businesses, newspapers, or other individuals who wish to buy time on a cable channel and then present their messages, sometimes with commercials included. Often these channels were shared (for example, by the local government and the schools), and sometimes there were several channels for the same type of organizations (for example, one channel for the local college, one for the public school systems, and one for private schools.)

Of course, not all the local programming planned by cable systems and local groups actually materialized, but the quality of access programming improved greatly from the early days when access consisted mainly of vanity TV.[43]

## Interactive Cable's Success and Stumbling

Interactive cable was highly touted in franchise applications. Once again promises were made that cable would perk the coffee, help the kids with homework, do the shopping, protect the home, and teach the handicapped. In some areas of the country a fair amount of interactive cable was actually undertaken. Cable signals that go from a central location to a home by wire also have the capability of going from the home to a central location either on the same wire or a different one. This enables cable to take on a two-way capacity and allows for interaction between the subscriber and whatever entities the cable system wishes to provide.

The pioneer and most publicized interactive system was Qube, which was operated by Warner-Amex cable. Qube was initiated in Warner's Columbus, Ohio, system in 1977 and later was incorporated in several of its other systems.[44] The basic element of the original interactive Qube was a small box with response buttons that enabled Qube subscribers to send an electronic signal to a bank of computers that could then analyze all the responses. An announcer's voice or a written message on the screen asked audience members to make a decision about some question such as who was most likely to be a presidential candidate, whether a particular congressional bill should be passed, what play a quarterback should call, whether an amateur act should be allowed to continue, or whether a city should proceed with a development plan. Audience members made their selections by pressing the appropriate button in multiple-choice fashion. Then a computer analyzed the responses and printed the percentage responding to each choice on the participant's screen.[45]

In 1981 Qube added to its interactive capability the ability for subscribers with computers to access a computer data bank to obtain a wide range of information such as consumer tips, weather, news, video games, and articles from magazines.[46]

Working toward fulfilling the promise that cable would help with the shopping, Times Mirror provided some of its subscribers with a shopper's channel that showed consumers various items and offered them goods at 20 to 40 percent below normal retail prices. This was possible because of a tie-in between Times Mirror and a product sales company, Comp-U-Card. This was not truly cable interactive because subscribers ordered items by using the telephone, but plans were considered for purchasing mechanisms utilizing the return capability of the cable.[47]

Another quasi-interactive service offered by some cable systems was pay per view whereby subscribers who paid an extra one-time-only sum could see one-time-only events such as championship boxing matches or special movies. Most pay per view plans asked subscribers to phone the cable company to order the event which would then be sent through the wire only to those homes requesting it and paying for it. Special addressable equipment was needed in order to service part of the cable system while keeping the event from entering the homes of people who did not want the pay per view event.[48]

Home security was another interactive area that cable entered. The two-way capability allowed for various burglar, fire, and medical alarm devices to be connected to the cable system. A computer in a central monitoring station could send a signal to each participating home about every ten seconds to see if everything was in order. If any of the doors, windows, smoke detectors, or other devices hooked to the system were not as they should be, a signal was sent back to the cable company monitoring station, which then notified the police.[49]

None of these interactive services were profitable, but proposals for them caught the attention of city councils and were often responsible, at least in part, for decisions regarding franchise awards.

### Consolidation and Retrenchment

The mid-80s saw a defoliation of the cable industry. Perhaps the promises had been too lavish and the anticipation too great, but the bloom was off the rose by 1985. The rate of growth of new subscribers leveled while the rate of disconnects from old subscribers increased.[50] Programming services consolidated and went out of business and those which were left sported much of the same type of play-it-safe programming that had traditionally been found on ABC, CBS, and NBC. Multiple System Owners (MSOs) drowned in red ink as they tried to live up to the promises they had made in regard to wiring big cities. Advertisers did not respond to the cable lure as quickly or profusely as expected. Even a Time, Inc. magazine about cable TV programming failed.[51]

Companies took actions to stem financial woes. Warner-Amex, for example, sold several of its cable systems to allow it to concentrate resources on a few cities. Storer and Times-Mirror traded several systems to make the systems each owned more geographically contiguous.[52]

In the area of pay programming, Showtime and The Movie Channel merged in 1983, with Viacom and Warner-Amex becoming the co-owners of both. The two separate pay services were maintained, at least for the time being.[53] HBO, the perennial leader, remained profitable but suffered a slowdown that led to changes in the executive suites. HBO's parent company, Time Inc., maintained its other pay service, Cinemax, although this venture was not profitable.[54] Playboy diversified by making its programming available to the cassette market and the subscription TV market,[55] and Bravo changed its programming emphasis and marketing strategy several times trying to find its niche.[56] In 1983, a pay service owned by Times Mirror, Spotlight, totally folded; and another, The Entertainment Channel, the cultural programming service owned by Rockefeller Center and RCA turned its programming over to the basic service, ARTS, which then became the Arts and Entertainment Channel.[57]

The pay services also began making exclusive deals with various movie studios to guarantee that certain films would appear only on one company's channels. Showtime/The Movie Channel, for example, made an agreement with Paramount that gave it exclusive pay TV rights to all theatrical features produced by that studio from 1983 through 1988. This meant that none of Paramount's films would be shown on HBO.[58] The pay services also began making their own exclusive movies and programs. Time, Inc. joined with CBS and Columbia to form a production company, Tri-Star, to produce movies for HBO.[59]

Pay cable services also suffered from piracy in the form of signal theft by people who had not paid for the programming. Some of this was within the cable systems where people figured out methods of obtaining the pay services by tapping off a neighbor's feed or altering technical devices so that the signals would enter their homes. In addition, an increasing number of people were buying their own satellite dishes and receiving the pay TV signals directly from the satellites. The cable industry figured its losses at $500 to $700 million annually because of service theft.[60] This led HBO to scramble its signals—a costly process.[61]

Problems occurred in the basic services also. The most highly touted failure was that of the cultural service CBS Cable which stopped programming after losing $50 million. The service had programmed ambitious high quality television, but did not receive sufficient financial support from either subscribers or advertisers. Because it had touted its service so aggressively, its demise was almost heralded by some of the cable companies who resented the encroachment of the broadcast networks into their cable business.[62]

Another well publicized coup occurred when Ted Turner's Cable News Network slew the giants, ABC and Westinghouse, by buying out their Satellite News Channel. This meant less competition for CNN which was then able to proceed on less tenuous financial footing because it did not have to compete for advertisers with SNC.[63]

Daytime and Cable Health Network merged to form Lifetime.[64] The children's service, Nickelodeon, began accepting commercials. CBN changed from a religious format to a family-oriented format that included very old reruns.[65] Getty Oil was bought out by Texaco which did not have any interest in maintaining the then unprofitable ESPN. After a brief flirtation with Ted Turner, ESPN was sold to ABC.[66]

There was at least one very bright spot on the basic programming level— MTV (Music TV). This Warner-Amex music video service aimed at teenagers became a sociological phenomenon and a financial success. Just years after its 1981 start-up date, it announced it was in the black primarily because of advertising support. As a result, several other music video services were quickly born. One of these, Cable Music Channel, launched by Ted Turner, had a short life span of less than a month.[67] Another, VH-1, was planned by MTV itself to appeal to an older audience.[68]

Activity on the local access and local origination channels also slowed during the mid 1980s. Cable systems which had promised truckloads of production equipment to local organizations tried to drag their feet on these obligations because they were so costly. Local programming departments which had included five or six employees dwindled down to a precious few. In some instances the director of marketing took over programming "on the side." Local programming did not by any means disappear, but it did not fill the multitudes of channels promised in many of the franchise agreements.[69]

In the area of interactive services, Warner-Amex had essentially killed its Qube system by 1984, mainly because it was not profitable.[70] No other interactive services rushed in to fill the void, so the whole interactive area remained stalled, at best.

Many proposals were made for pay-per-view but, although the field looked riper, little had proven financially successful.[71]

By 1985 cable TV was far from gasping its last breath, but predictions for the future were cautious and the wunderkind aspects of the gold rush days seemed like ancient history.

## The Spirit of Deregulation

Cable TV, like broadcasting, has been affected by the deregulatory mood of government. In 1980 the FCC abolished both the syndicated exclusivity and distant signal importation rulings, opening the way for cable systems to import as many stations as they wished and to play the programming of those stations without having to worry about whether the same programming would be on a local TV station.[72] Broadcasters, motion picture producers, and sports interests, concerned about the effects of this deregulation on their businesses, appealed the FCC ruling, but the Supreme Court ruled in favor of the FCC action and, therefore, in favor of the cable TV interests.[73]

Cable TV looked like it was home free on the two long standing issues of distant signal importation and syndicated exclusivity, but in 1982 the issue of copyright interceded. The Copyright Tribunal, acting under its right to adjust fees, ordered cable systems to pay 3.75% of their gross receipts for each distant signal imported that was above the number of distant signals allowed under the 1972 rulings. This ruling had set up a complicated formula for distant signal importation that depended on the size of the cable system, the size of the market, and the number of stations in the market. The only systems affected were fairly large ones in densely populated areas, but these systems did serve about 10,000,000 people. The cable companies, of course, appealed the decision, but the courts ruled in favor of the Copyright Tribunal. This action pleased program producers because it meant more money for them if the cable companies paid the fee. Some of the cable systems affected, however, dropped distant signals, particularly the super stations, so that they did not have to pay what they felt was an exorbitant percentage. The net result favored the producers, though, for in 1983, the first year the 3.75% fee was in effect, cable operators paid $25 million more to the Copyright Tribunal than they did in 1982 for a total of $69.2 million.[74]

A major piece of deregulation legislation passed by Congress in 1984 delineated the role of cable systems and local governments. The bill was heavily lobbied by the League of Cities and by cable interests, but both seemed fairly happy with the results. The main positive change for cable operators was that cities could no longer regulate the rates which cable systems charge their customers. Cities could still oversee public, educational, and government access programming requirements, and they could collect up to 5% of the cable companies gross operating revenues as a franchising fee.[75]

Must-carries surfaced again, and, in 1985, they were abolished. This was due largely to arguments put forth by cable companies that did not have enough channel capacity to carry all the satellite services they wanted to carry. They argued that local stations can be received over the airwaves, so cablecasting them constitutes an unnecessary duplication of services to the consumer. Local stations, of course, did not agree because they wanted to be carried on the cable systems, but they lost the battle and the FCC discontinued the must-carry rule. Cable companies, in general, continued to carry the popular, greatly watched stations, but some began dropping the less popular outlets, particularly UHF, religious, and public TV stations.[76]

A new meaning has also surfaced for must-carry. TV stations and cable systems are now arguing whether or not TV stereo broadcast by a station must be carried as stereo by the cable system or whether it can be changed to mono. Likewise, questions have been raised regarding other services which broadcasters provide such as subcarrier offerings and teletext. These are issues which will continue to be under review.[77]

Overall the role of government in general and the FCC in particular in the regulation of cable TV has been muddled. Because the FCC controls the airwaves and cable transmission is over wire, its jurisdiction is questionable. Although the government has made some decisions concerning such issues as distant signal importation, syndicated exclusivity, must-carries, and copyright, these issues will continue to be argued and altered.

## Issues

The number of issues revolving around the cable TV area is enormous. Part of this is due to the rapid changes in the industry. The pace is so fast that many of the ramifications of actions cannot be thoroughly thought out. The time needed to solve problems is often not available in the hectic crunch to churn out programming, string cable, sign up subscribers, and solve financial problems.

### Internal Problems
Some of the issues facing cable are internal to the industry. For example, the old perennial problem of pole attachments still plagues the industry. Disconcerting attempts to work in harmony with the telephone companies, so that cable could

be strung on telephone poles, are almost as old as the cable TV industry. The problem still exists and, no doubt, will continue. The phone companies can always claim to be understaffed to the degree that they can not handle the cable company requests promptly. While they spend months trying to decide how to rearrange wires to enable the cable company to place cables on the poles, the cable TV general managers are losing sleep over the loss of dollars attributable to slow wiring and hence, slow hook-ups of paying subscribers.

The intertwined problems of copyrights, must-carries, syndicated exclusivity, and distant imports keep cable TV lawyers on their toes. In many ways these issues are totally unrelated and yet sometimes they become intertwined for political reasons. If broadcast interests give a little by allowing unlimited distant importations of stations, then cable interests give a little by agreeing to continue the cumbersome practice of blacking out programs that are also on local channels. All parties interested in these issues, movie production companies, unions, writers, sports packagers, musicians, and distribution companies are constantly jockeying for positions that will better their lot in life. For this reason these issues will probably continue to be compromised and intertwined.

Another internal squabble likely to erupt more in the future, than it has in the past, is the issue of who pays how much to whom for programming. As it now stands, most cable networks charge cable systems that show their programs. But if the networks become more successful in obtaining advertisements, the cable systems will probably feel that they do not need to bear this burden. They will begin pushing for a system similar to what broadcast networks now use with their affiliated stations. The networks actually pay the stations to air the programs because the networks keep the advertising money. Advertisers are charged according to the size of the audience reached, so the networks want as many affiliates as possible to show the programs. In that way they can charge a higher fee to the advertisers. Sometimes, in addition to paying the stations to air the programs, the networks leave a few advertising spots blank so that the local stations can sell ads for those spots and keep the revenue.

This type of set-up would certainly work in the cable world, too. In fact, it might work more appropriately because cable networks compete with each other to be given channel space on any particular cable system. The cable systems would be able to bargain with the networks for the best deal in terms of both financial remuneration and open advertising time. No elements of the cable industry are making enough profit for this to become a major issue as yet, but the potential is certainly there.

### Economic Reality
Although cable is by no means down and out, it has had to face some hard economic realities. During the gold rush days, there appeared to be unlimited possibilities for profit in the cable field. But that was not to be. So many companies tried to cash in on cable that the market became glutted. A shakedown for survival of the fittest took place leading all of cable TV toward economic reality.

Along the way companies, individuals, and concepts suffered. Large and small companies faced bankruptcy or severely damaged balance sheets. Individuals lost their jobs as cable systems and satellite services cut back on staff in order to cut back on expenses. Ideas that might have been workable had they had more development time and/or more capital investment fell by the wayside.

Particularly hard hit was the concept of culture on cable TV. With the demise of CBS Cable and The Entertainment Channel, the possibilities for advertising supported or subscriber supported culture seemed bleak. Cultural programming may not be able to survive without government subsidy. This has certainly been the case in other countries, and public broadcasting in the United States did not achieve prominence until the government established a support system.

None of the areas of cable programming—pay, basic, local, or interactive—were immune to economic cutbacks. Pay networks consolidated and groped for solutions such as pay-per-view and unusual specials. Likewise basic services consolidated and changed programming concepts from broad to narrow and back again in an attempt to find a formula that would appeal to advertisers. Cable systems cut back on local programming, failing to activate many of the promised access channels and providing less and less support, in both equipment and personnel, to the local programming channels which remained. Interactive services folded and were not replaced.

All of this meant that cable failed to keep many of its promises—to its stockholders, to city governments, and to the public at large. Now that the shakeout has occurred, however, cable TV shows signs of recovery. Perhaps if it travels along a less flamboyant road, it will achieve a comfortable niche, both socially and economically.

## Ownership

No longer is cable TV run by small time entrepreneurs; it is now big business. The money, the decisions, and the power come from large corporations. This tends to give the industry a degree of financial security, but it also can lead to abuses because power corrupts and absolute power corrupts absolutely. Often decisions that affect local areas and local mores are handled at a corporate level far from the scene. Corporate giants are often not sympathetic to experimentation and many of the fresh, cocky ideas, which could be tried in cable in its more youthful days, are no longer possible.

The varying ownership interests show signs of history repeating itself. For many years the major film production companies not only produced films but also distributed and exhibited them. In other words film companies owned movie theaters. In 1950 the Supreme Court ruled that this type of vertical integration in companies was restraint of trade and ordered film companies to divest themselves of either production, distribution, or exhibition. Most film companies sold their movie theaters and continued to produce and distribute films. The effect of this was a collapse of the monopoly that the major film production companies had on the film market for some thirty years. Small independent film makers now had a greater chance of having their films exhibited because the major companies did not own the theaters and could not confine the product exhibited to that which they had produced.

But a situation similar to that which existed prior to the 1950 ruling has developed in cable TV because companies are becoming involved in production, distribution, and exhibition of cable programs. Time-Life owns HBO, which distributes movies by satellite. It has also joined with CBS and Columbia to form Tri-Star, a company which produces movies. In addition, Time-Life owns ATC, one of the top multiple system owners in cable. Through this arrangement Time-Life can become involved in production (Tri-Star), distribution (HBO), and exhibition (ATC). This does not seem to differ from the film company arrangements prior to 1950 and may bear careful watching in this regard.

Another ownership topic to cause controversy is cross ownership. This term has its roots in early broadcasting when city newspapers also tried to own city radio and TV stations. The FCC feared that one company would have too great a hold on the dissemination of information if it controlled newspaper, radio, and TV so established rules (which it changed often) regarding the extent to which this cross ownership could occur. In general, the FCC tried to prevent monopoly within any one city. It did not object to a company owning a newspaper in one city and a TV station in a far distant city, but it did object to multiple ownership in the same area, especially if that area had a limited number of media outlets. As cable systems began to flourish and be taken over by large MSOs, some of whom were in the broadcasting and/or newspaper business, they, too, were considered under the cross ownership rules. As a result, when Westinghouse merged with Teleprompter, Westinghouse had to sell either its broadcast properties or its newly acquired cable properties in some areas where the two coincided. There are some who argue that cross ownership rules should be abolished because the number of different media sources people now have access to has grown enormously and that control of several of them by one company will not prove dangerous.

In a similar vein, broadcasters have only been allowed to own one TV station (channel) per city. Now cable TV systems have appeared, some of which program over 100 channels. Although these channels are not all owned by the cable system, they are controlled by it. If a cable system did not agree with the ethnic programming of Black Entertainment Television, for example, it could simply refuse to show that channel. A cable operator potentially has greater power to control what the viewers see than the broadcast operators who can only control the content of one channel.

Ownership configurations are changing so rapidly in cable that rules regarding them cannot keep up, but the industry itself and the public watchdogs should keep an eye on the trends so that unjust malpractices do not affect the business structure or the consumers.

## Relationship to Government

Cable TV has always had a rather unsettling relationship with government entities, and this will no doubt continue. Historically, various levels of government have been hesitant to regulate cable, leading it to grow in a rather topsy-turvy manner. It has had to fight for every piece of favorable legislation it has received and has not been privy to guardian angels to support its concept on the legislative front. Its adversarial relationship with other entertainment organizations will no doubt continue and will lead it into further difficulties in governmental areas.

Local governments are the ones which deal most directly with cable TV, primarily through franchising and refranchising. Much of the local government power has been taken away because cable TV systems are now free to set their own rates without approval from city governments. This could lead to rate inflation damaging to consumers' pocketbooks. Subscribers, however, always have the option of disconnecting the service, so if the cable companies boost their charges too high, they will lose customers and, therefore, income.

The future franchising process is less likely to involve the promises and corruption which accompanied the franchising of the late 1970s and early 1980s. Fewer companies are interested in obtaining franchises because they have not proven to be as lucrative as predicted. Most franchises are now awarded, so the main actions to be taken involve refranchising. Because most franchises last for fifteen years, refranchising is not a process which is likely to cause a great deal of trauma in the near future.

The main conflicts arising between cities and cable companies at present involve unkept promises. In some instances cable companies receive franchises and then proceed to build the system at a very slow pace. This irritates the city which can take away the franchise. When a company is slow to build, the usual cause is lack of money, sometimes to the degree of bankruptcy. What often happens is that the cable company which received the franchise is forced to sell its system to another company. The city is then in a position of dealing with a company it did not select. The new company may try to change the rules so it does not have to fulfill all the promises made by the original company. This is likely to cause lengthy and sometimes bitter negotiations.

Conflicts are bound to continue between cable TV and government, but that is common between any regulators and regulatees.

## Advertising

Many people originally subscribed to cable TV because they were tired of all the commercials on network TV and were willing to pay to watch programming that did not contain the bothersome ads. Although the pay services have managed, so far, to remain advertising free, the basic services have not.

In fact, many of these services are primarily advertising based and are looking more and more like the networks every day as they do, indeed, attract advertisers. Both the quantity and quality of commercials is becoming the same. There are no rules limiting the number of ads cable channels can run, so they are taking all comers. The ads being placed on cable are, for the most part, the exact same ones that appear on broadcast TV and have been condemned for being inane, uninformative, tasteless, and misleading.

Cable practitioners talk about "informercials" but very little has been actually undertaken in this area. Informercials are intended to give actual information about products in a length of time that exceeds the thirty seconds of most broadcast commercials. Instead of being cute and fast paced, they would actually explain the product, demonstrate how it should be used, and tell its advantages, its limits, and perhaps even its shortcomings. A few informercials have been produced, but they seem to be extended hard sells rather than actual information.

Whether cable actually begins programming informercials or whether it merely continues to program traditional commercials, it still short changes the viewer. Cable is no longer an escape from advertising. Instead of merely having to endure commercials, the viewer must now pay extra and also watch the ads.

Advertising is bound to continue in cable because it represents a welcome source of income. Only complaints and action on the part of cable subscribers can keep the cable services from becoming proliferated with ads, if, indeed, the advertisers wish to place their ads on cable.

### Effects on the Poor

Because a fee must be paid to receive cable TV, some members of society cannot afford it. These people may be the ones who most need it because they are the economically poor who are also the information poor.

Traditionally, cable companies have wired poor areas of the city after they wire the more affluent areas because they are able to achieve a higher penetration in the richer areas and thus recoup their costs more quickly. Many franchise agreements now require cable companies to begin building in various areas simultaneously, but that does not help the poor pay for the cable services. Rural areas are often not wired, either, because the cost of stringing all the cable in areas of sparse population far outweighs the income that might be generated.

The poor have broadcast TV and those in rural areas have relayed broadcast signals from which to obtain information about the world around them. However, some programs are shown only on cable and only to those who pay for them. Under these circumstances the rich can become more informationally rich and the poor will be more informationally poor. This could mean that various factions of society will grow further apart. People who begin with an economic advantage will widen that advantage.

The counter to that argument is that the poor will, indeed, subscribe to cable TV. Homes that cannot afford telephones or bathtubs have TV sets. If the poor can find a way to afford TV sets, they can find a way to add a modest cost for cable programming.

## National Programming

There are many issues that revolve around the structure and content of the national cable TV networks, but none arouses greater controversy than the subject of sexually explicit programming. Cable networks have taken the point of view that people pay to receive this programming and are totally aware of what they have orderd. They should have the freedom to see sexually oriented material in the privacy of their homes if they so desire.

And people do desire. When these services are offered on cable, they usually achieve over 50 percent penetration and can be sold for about nine dollars a month per subscriber.

But bringing the services onto a cable system at all is highly controversial. Many cable systems that are owned by multiple system owners will not touch these so-called cable-porn services because their parent companies are attempting to receive franchises in other parts of the country. Being in any way associated with a system that offers this fare is very detrimental in the franchising process.

In addition, community groups almost inevitably fight the introduction of such channels, and the individuals who appear to be responsible for wanting to institute them receive a flood of bad publicity. The people who fight the channels do so primarily on the basis that introducing such material into the community will negatively affect the quality of life in the community and may be particularly disastrous to young people.

Interestingly, once the channels overcome these hurdles and are introduced, they receive far less criticism. Perhaps because of the sensitivity of the issue, such services do not program highly erotic materials. The movies are generally R-rated and can be seen at drive-in theaters. The services program only in the late night and early morning hours. The Playboy Channel is patterned after the magazine, which has a reputation for somewhat sophisticated adult material.

The films that fill the movie services such as Eros and Private Screenings are inexpensive to make. Because a large number of people wish to subscribe and are willing to pay a fairly steep price, a great deal of money can be made in the business. But the production of the films also arouses ire.

The business of producing what are generally considered pornographic movies is considered a sleasy business. People (actresses especially) who perform in these films often become type-cast and cannot find work in other types of more socially acceptable films.

The criticism of both sex and violence on cable TV goes beyond the channels that are clearly designated as adult channels to other pay services such as HBO, Showtime, and the Movie Channel. These channels show a variety of material from G-rated to R-rated. Parents who wish to make sure that their children do not see films that they feel are undesirable must keep very close track of what is showing when. Generally, these services do not show R movies with high degrees of sex and violence during the day, but certainly the choice of times for some of the films could be considered questionable. In fact, questions can arise as to whether these services should show R-rated material at all, whether it be movies, stand-up comics, or specials. The service is bought as a package for a monthly fee. Unless people very carefully scrutinize the program guides, which detail the channel offerings, they can find they have brought into their home programming they do not want.

Programmable lock boxes are available on a limited basis. Parents can buy these and program in specific times of the day when they do not want a particular channel shown. However, this involves an investment of both time and money on the part of the parents and may negate the value of the channel.

Overall, the controversy of sexually explicit programming is a complicated one. It involves such issues as whether or not society should accept and promote this type of programming at all, the freedom of individuals to choose what they wish to see, the tastefulness of content and presentation, the time of showing, and the packaging of the material within other types of programming. These are not easy issues, and they tend to be treated differently in various parts of the country and at different times in our evolving history.

An entirely different issue related to national cable programming revolves around the number of programming services available. Survival of the fittest has eliminated some of them, but there may still be more than are socially beneficial. While programming on the three broadcast networks may be considered inane at times, it still does tend to unify the nation. The viewing of the programming has created a common denominator effect because the programming is discussed nationwide by all classes and categories of people. Broadcast TV may be the only truly mass media left. Radio audiences are fragmented into sub groups with similar music or programming interests, and magazines have been designed to appeal to small interest groups. If TV audiences, too, become subdivided into small groups watching numerous channels, the unifying effect of TV viewing will be lost.

The affect of cable on movie theaters may also be great. Broadcast TV severely reduced the number of people who went out to movies, and cable, given its emphasis on movies, may reduce this even further. Pay-TV services have the capability of eventually bringing in more money for movie producers than theater distribution does. This means producers will be willing to give the pay services the films at the same time or earlier than they are given to the theaters. If this happens, the theaters will lose what little edge they now have for attracting audiences. Theaters are attended primarily by the young who wish, and need, some form of entertainment away from their homes. If theaters become financially unviable because cable takes their product and their profit, then the social patterns of the young will undoubtedly alter in years to come. Movie theaters could go the way of vaudeville theaters, but movies themselves are likely to be more in demand than they are now. The public will always be hungry for entertainment, so although the exhibition aspects of the movie business may change, the production aspect will thrive.

Other aspects of national programming are coming under scrutiny. Several of the cable channels have bought series cancelled by the broadcast networks. This may mean these services are heading toward series programming of the same form and content as broadcast TV. The pay services have also gained exclusive rights to some of the movies they show in order to gain a competitive edge on the other pay services. These two trends taken together could lead to pay networks that are similar in content and form to the present free networks. Previously most of the movies shown on one pay system were shown on the others during the same month, but the schedules of the various channels were not exactly the same. The main reason for viewers to subscribe to more than one pay service was to increase the flexibility of when the various movies could be seen. But if each service develops its own exclusive series and programs, then in order to obtain the equivalent of NBC, CBS, and ABC, a viewer would need to subscribe to three pay services—obviously, a hefty tab.

National TV programming has always attracted national interest and national criticism. Apparently, cable programming is no exception.

**Local Programming**

Some cable operators look upon local programming as a chance to serve the community in such a way that they will gain additional subscribers. Others look upon it as a burden that drains resources unnecessarily because no one watches it anyhow.

Local programming, both origination and access, exists mainly because of franchise agreements, but companies that take these agreements seriously have made excellent advances in the quality of such programming. Still there are many who feel that cable systems should not be required to undertake any form of local programming.

The area of local origination is embraced more readily by cable systems than the idea of access, mainly because systems see more of a chance to gain revenue through advertising with l.o. These advertising dollars, however, often come at the expense of local radio and newspapers, which charge approximately the same rates. Both of these media would prefer that cable not develop its local advertising based programming.

The area of public access can cause headaches for cable operators because, although the programs are cablecast over their system, they do not directly control the programming content. When material of questionable decency appears on a system, the cable operator is blamed even though it did not produce the program. The amount of control that a cable system has over access programming is nebulous. Some systems establish advisory boards made up of members of the community. These people draw up guidelines for the access channels, and the cable system employees then enforce them. In this way the cable company is not censoring but is following the dictates of the community. Other cable systems set up technical criteria that cable programs must meet and hope that these will keep the truly unacceptable programs out. Actually, most access programs are produced with admirable intent and contain worthwhile content. Often they are amateurish in production values because the people producing them are novices, but even that is improving because cable companies are conducting workshops to teach citizens the rudiments of TV production.

Leased access is another thorny topic, and, once again, many cable operators feel they should not need to offer this service. In one version of leased access a particular company leases an entire channel from the cable company and uses it as it sees fit. A newspaper, for example, may lease a channel to present a headline service. The cable company feels that even though it receives revenue from such an arrangement, it is giving up one of its channels that could, perhaps, be earning more revenue if used in another manner, such as for another pay movie channel. Other leased channels are available to any commercial groups who wish to pay to present messages on a one-time or series basis. These have some of the same problems of quality control that public access channels have, but while public access is generally non-profit for the cable company, leased access is designed to make money.

Interconnects are another topic discussed more in relation to local programming than national programming. Many cable franchises cover fairly small geographic areas, so programs produced locally do not even have the potential to be seen by a large audience. If a number of different cable systems are interconnected, then the local programming could be seen by a larger group, perhaps a group as large as can generally see local broadcast station programming.

Educational and government groups are often particularly eager for interconnects if their areas of jurisdiction do not coincide with those of the cable franchise area. For example, a school district may produce a series to be used on an educational access channel by all third grades in the district. But if some of its elementary schools lie outside the area that the cable company has wired, then those schools cannot receive the programming. Likewise, a police department might want to present material on neighborhood watch only to find that part of the "neighborhood" cannot receive the program because it is out of the franchise area.

But interconnecting systems is not all that easy. Even if the political problems of one company working with another can be overcome, the technical problems associated with physically connecting systems are difficult, especially if one system is an old one with only twelve channels, and the other is a new one with fifty-four channels. Some interconnects do exist, especially where the primary program material is advertising supported local sports, but interconnection is an issue that will receive more attention as time progresses.

Local programming of various forms is fairly well engrained in many cable systems now and will probably continue to exist whether there are franchise requirements for it or not. Admittedly, the audience viewing this programming is not large but may increase as cable itself achieves greater penetration and as the quality and significance of local programming increases.

## Interactive Services

Cable TV has a great deal of competition in its interactive role, and actually, much of what began as cable interactive has now come under the category of videotext. The videotext area is of interest to the phone companies, newspapers, broadcasters, and others. Cable is attempting to become the main purveyor of interactive services, but its role at present is tenuous.

Aside from the videotext, there are those both in and out of cable who wonder if home security is really a business cable TV should be involved with. The fact that wires can come back from a subscriber's home to the cable location makes it physically possible for cable to serve the security function. But, people in the business are not acquainted with the police-like functions of security work and are worried about the risk and liability they might incur if a home they have wired is robbed because some facet of the security watch did not work. What some systems have done is subcontract the security function to a company regularly engaged in home security and in that way provide the service without administering it.

Addressability is a fashionable idea within cable interactive services. It refers to the ability of a cable company to directly change or contact any particular TV set that it has wired without actually going to the home and touching the set. A device in the cable company office can "addresss" any particular set and give it instructions. In this way it can program the specific channels any set should receive and turn off channels when a bill is not paid. The issue here is addressability versus privacy. Any device controlling something as personal as TV viewing smacks a bit of "Big Brother is watching you."

More of the issues involved with interactive services will be discussed in the videotext chapter, but cable TV will be in there competing to provide the services and solve the problems.

## Piracy

Piracy of cable TV signals has become an important issue within the cable industry. It is being attacked on two fronts—through subscribers who are receiving signals through unauthorized wiring configurations, and through people who have rooftop satellite dishes.

People receive cable TV signals in a number of different ways without paying for them. They can tap off a neighbor's wire or they can remove traps which have been inserted in their cable hookup to prevent them from receiving certain pay signals. This, of course, means a loss of income for the cable company. The cable industry has mounted a campaign to stop such signal theft, but it has not been overly successful. People who are otherwise law-abiding citizens feel little guilt about bringing cable TV into their homes without paying for it. They have been conditioned to think TV is free, and therefore, do not feel they are stealing.

A great deal of hardware has been developed to prevent signal theft, but it is expensive for cable systems to install. If the signal theft is low, the cost of the equipment may not justify itself. Older systems are more likely to be subject to signal theft than newer ones because newer ones are equipped with the new signal theft equipment. Many of the newer systems are interactive. When signals flow back to the headend, the cable company can tell if someone who has not paid for it is receiving a signal. The rate of signal theft in cable TV is high enough to cause great concern in the industry.

An even more complicated issue is the reception of cable network signals by people with satellite dishes. These people made heavy investments in the dishes in order to receive the satellite signals. They feel they have paid by buying the dish. The cable companies, however, do not feel this way. They are not receiving any money from these people for the programming. In some instances, people with dishes live in areas also served by cable. If they removed their dishes and subscribed to cable, the cable systems and networks would be receiving income from them.

The cable industry's defense against this problem has been to scramble signals. The satellite dish owners still receive the signals, but in a scrambled, incoherent form. Scrambling is expensive, however, so once again the cable companies are paying a hefty price for signal security. Although there are plans afoot to sell the scrambled signals to satellite dish owners and provide them with descrambling devices, these plans are not yet finalized. Selling to these widely scattered individuals will involve a type of marketing which the cable networks have not yet undertaken.

The whole issue of piracy is controversial because it is not theft in the traditional sense. The product disappears as soon as it is "stolen." The signals flash on the TV set and are gone. Massive prosecutions of people who are receiving cable TV signals in a manner that the cable systems have not authorized are not likely. And yet, if everyone received the cable signals for free, there would be no cable TV.

## Relationships with Broadcasters

Cable TV is growing up during a period when deregulation rather than regulation is the predominant political philosophy. Much of broadcasting's journey to maturity took place during the 1960s when regulation was in vogue. Broadcasting, therefore, looks with both envy and anger upon the cable industry, which needs to think about renewal of the franchise only every fifteen years and is able to offer a large multitude of programming services.

Many broadcasters harbor the long-standing animosities, which broadcasters and cablecasters have tormented each other with for years, over such issues as copyright and distant imports. But as old-line broadcasting companies make inroads in the cable business, the distinction between cablecaster and broadcaster is becoming blurred and the animosities are disappearing. Still, broadcasters worry that cable may usurp their popular programming, and feel they should be protected from such erosions because they provide free services to the people while cable is only for those who can pay.

On the other hand, old-line cablecasters resent the intrusion of the powerful broadcasting companies into their business. They are the ones, who like the little red hen, struggled for years to grind the wheat and bake the bread. Now that cable is important, the former enemies are very eager to come and partake of the bread. Ted Turner, for example, who pioneered in the cable area, lambasts the networks for their Johnny-Come-Lately entrance into the cable business.

The cable TV industry has many more issues surrounding it than the other media usually considered as new technologies. Part of this is due to the fact that cable has been around longer than most other new technologies and has a history of conflict and struggle to draw from. Also, cable TV has become visible and is finding that the price of success is constant criticism and attack. Cable is no longer the underdog. It must fight its own battles and stave off its own competitors while at the same time trying to please its customers and make an old-fashioned American buck.

# subscription tv

## Description

Many people confuse cable TV and subscription TV (STV) because commercial free movies are the primary programming fare of subscription TV, just as they are the primary fare of the most publicized aspect of cable TV, the pay movie channels. To confuse the issue further, both are often referred to as pay TV. However, the two are very different in terms of transmission.

### Over-the-Air Scrambled Transmission
While cable TV has a rather complicated delivery method that involves satellites, microwave dishes, and coaxial cable, the transmission system for subscription TV is quite simple. It utilizes air waves just as the regular broadcast stations do. The only difference in transmission between a network affiliated station and a subscription TV service is that the latter scrambles its signal before it is sent out from the antenna.

Many different methods of scrambling exist, but basically they all rearrange elements of the TV signal so that it is unrecognizable to a regular TV set. Some scramblers reverse polarity so that blacks are whites and visa versa and then also reverse horizontal and vertical impulses, in essence, turning the picture inside out or jumbling it like a well-mixed jigsaw puzzle. More sophisticated scramblers constantly change how the scrambling is accomplished so that at one moment horizontals may be shifted one direction and the next moment they may be shifted another direction.

Anyone who turns on a TV station that is scrambling its signal will receive an image on the screen, but it will be a very confused, incoherent image. In order to turn the image into a coherent picture, a decoder (also known as a descrambler) is needed. This is usually in the form of an inconspicuous box, which can sit on top of the TV set. This decoder takes the scrambled picture and rearranges its elements so that it is once again a proper picture. For example, if polarity was reversed at the transmitter end, the polarity will be reversed back the way it should be. For scrambling methods that change frequently, a special signal is sent to the decoder box to tell it when to change the method of descrambling.

**Figure 6.1.** STV configuration.

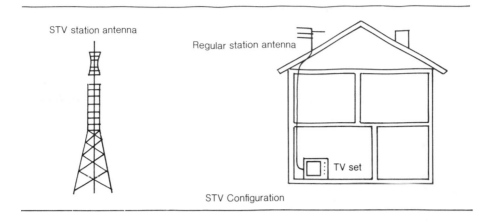

STV station antenna

Regular station antenna

TV set

STV Configuration

## Addressability

Some subscription TV systems are sophisticated enough that they can give special instructions to each decoder box in their area from the office of the STV company. This is accomplished by sending special digital codes, which give commands to individual boxes. In this way STV can offer programs, which are descrambled, in only some of the homes. The ability to contact individual boxes is known as addressability.

This addressability can be used when someone doesn't pay a bill. The company can turn off the service from the STV facility rather than sending a repairman to disconnect it. It can also be used for pay-per-view when subscription TV services offer special attractive events such as boxing matches or rock concerts. STV companies can send a special signal to the decoder boxes of all the people who paid extra for the program so that it is descrambled in those particular homes.

A few systems have experimented with addressability to make all their programs pay-per-view. In other words, viewers would only be charged for the programs they actually watched. This can be accomplished by having viewers call and order programs or by having decoders keep track of programs watched for billing purposes. However, this requires an enormous amount of publicity on the part of the STV operator in order to make viewers aware of all programming available, and it requires more staff to handle individual requests and billings.

Another use for addressability is to allow certain subscribers to have some particular programs on a regular basis. For example, some STVs program sexually explicit films late at night. Those people who wish to receive the films can be addressed to do so while others will see scrambled material when these films are shown.

Some companies just as a general practice address each decoder box once a day to make sure it is operating properly because they have found that this cuts down on service calls. Some brands of decoder boxes are dependent upon being plugged into electricity constantly. If there is a power outage or someone unplugs a decoder to plug in a vacuum cleaner, the decoder will not reactivate itself when it once again receives power. By addressing the boxes each day, the STV company can reactivate the non-functioning decoders and correct these minor problems.

## Single Channel Local Operation

Unlike cable TV, subscription TV, consists of only one channel. This one channel is a regular broadcast channel, which was long ago allocated by the FCC to the particular city. Most subscription TV channels are in the UHF rather than the VHF band because the idea of subscription took hold long after all the VHF channels were being operated successfully on a commercial basis. However, many UHF channels had died, were struggling to survive, or had never been activated.

Subscription TV companies formed partnerships with the UHF channel owners so that each would program part of the day. Usually, the conventional programming was unscrambled during the day, and the scrambled pay programming came on at night.

The person who subscribes to subscription TV receives one channel's worth of movies and specials but does not receive multiple channels of movies or any of the basic or local programming services that accompany cable TV. This has been a real detriment to subscription TV because in order to make ends meet, STV must charge as much for one channel as cable TV charges for a whole compliment of channels.

Because STV utilizes regular broadcast channels, its coverage only encompasses that of a local station. Pay-cable programs are placed on satellites to be received and retransmitted by cable systems throughout the country, but subscription TV programs travel only as far as the broadcast signal will carry.[1]

## Programming

Subscription TV acquires most of its programming from the same place the pay-cable services do—the major motion picture production and distribution companies. Usually, STV is showing the same films as cable at the same time. This is due to the fact that the motion picture companies have set times when they release films to pay media. Usually, films are shown first in theaters and later are released to the pay TV market (both pay cable and STV) followed by release to network TV and then local TV stations. At each step the price for the right to show the film is negotiated, and prices can vary greatly depending on the movie, the potential audience size, and the skill of the negotiator.

The amount of time between theatrical release and pay-TV release varies from film to film and is referred to as a "window." For example, a film that is sold to pay TV six months after it finished in theaters is said to have a six month window. The real blockbuster movies will be placed on the shelf after their first theatrical release so that they can be released again, perhaps a year or two later, to recoup another round of profits from the movie theaters. Four or five years may pass before these movies are released to the TV market, or they may never be released at all. Movies with less box-office appeal may appear within weeks after theatrical release.

Movies are also produced that do not go to the theaters at all but receive their showings solely on TV. The broadcast networks have for many years aired made-for-TV movies. A few of these have been of such quality that they tour the theatrical circuit after being on TV, but most made-for-TV movies are of lower quality than those made for theaters. Some of the companies involved with pay TV are now producing made-for-pay-TV movies.

But most movies that are shown on STV come from the theatrical circuit. When a film company decides that time has come to release a particular film or films to the pay market, it will generally make the product available to all such companies who are willing to pay. Occasionally, through some special negotiations, one company will be given exclusive rights to show a particular film.

Representatives from each pay-cable or subscription TV entity decide which films they want to buy and when they want to program them. Copies of the film, usually transferred to videotape, are then delivered to each company for them to use for transmission.

In addition to general theatrical movies, some STV systems program sexually explicit movies, either R-rated or unrated, during the late evening and early morning hours. These are obtained from the groups that produce and/or distribute them and usually are available as soon as they are produced.

Some subscription TV stations depend on movies alone for their programming. Others engage in their own original programming of high quality events. The most successful ventures in this area involve obtaining the rights to broadcast local athletic teams. If big league baseball, football, or basketball teams agree to allow their games on subscription TV, then the STV has a major selling point.

Teams are more likely to allow STV coverage than broadcast TV coverage because the former encompasses a fairly small audience and is not likely to reduce significantly the number of people attending the games. Teams are reluctant to allow home games on regular free TV available to everyone because of the fear of diminishing gate receipts, but STV operators have been able to negotiate broadcast rights.

In addition to sports, STVs sometimes produce specials or limited series, which they show to their subscribers as part of the regular service. Sometimes these are in the form of experimental or special effects programming, such as 3-D movies for which subscribers must wear special glasses.

Most special events, however, are shown on a pay-per-view basis whereby the viewers pay a special fee for the one-time showing of a particular event. Boxing matches have been the most common and most successful of these ventures followed by rock concerts. Up to 50 percent of the regular subscribers request this special programming, but the percentage is usually closer to 30 or 40 percent.[2]

## Sales

Like cable TV, STV companies must sell their services directly to the consumer. This involves large advertising and promotion campaigns as well as door-to-door selling. Because only one channel is involved, the concept of STV is easier to explain than the concept of multi-channeled cable TV.

Subscription TV is available to subscribers quickly and in a wide area. As soon as an STV operation turns on, it can be received by everyone within range of the station. No long wait to cable an area and no franchise boundaries. Mass promotion campaigns over an entire city can be activated to build strong citizen awareness.

Subscribers pay a one-time installation fee and then a regular monthly fee. For this they receive uncut, uninterrupted, commercial-free movies and perhaps local sporting events and some other special programming. If sexually explicit movies are offered, they may require an additional amount to the monthly fee, or they may be included in the basic price. When a pay-per-view event comes along, the subscriber calls the STV company and orders it. The subscriber's decoder is addressed to receive that program and the amount for the show is added to the viewer's bill. The decoder boxes, themselves, may be leased or purchased from the STV company.

When a subscriber wishes to discontinue the service, he or she calls the STV company, and a representative disconnects the decoder box. If the box was leased, the representative brings it back to the company. Disconnects are a hefty problem for subscription TV companies. Some initial connections are made because people want to watch a special event, such as a world heavyweight championship bout, and once the bout is over they disconnect the service. A certain percentage of people continue as subscribers, however, so the specials actually draw people for the regular service.

The more damaging disconnects come from those people who are disenchanted with the movie programming. There are periods when there are few, if any, significant films to show, and that is when most disconnects occur. The reason for this problem relates, in part, to the theatrical film pattern. High budget, big star theatrical films are planned so that they are released either at Christmas or the beginning of summer, the two periods when the primary theater audience, young people, is most likely to attend the movies. If the released films are of an even quality, then they are all likely to come up for rerelease to the pay TV market at about the same time, several months down the road. With all the truly popular movies becoming available at approximately the same time, there are many months when hardly any truly worthwhile films become available. STV systems could hold back and show some films at a later time, but then their competitors in the pay cable field would appear to have a jump on the films. During the "dry" months, STVs receive many requests for disconnects, more so than the cable TVs, which are offering a variety of services.[3]

## Regulation

Because STV is an over-the-air service, it is regulated by the FCC just as all other broadcast services are. For many years the rules governing STV were designed primarily to protect conventional broadcasters and to thwart the STV competition. For example, rules prevented subscription services from bidding on programs broadcasters wanted. Also, STV was not allowed in cities that did not have at least four commercial TV channels so that it could not unduly harm small stations in small towns by taking away audience members.

However, in the spirit of deregulation, the FCC removed most regulations regarding STV, and left it free to grow. Unfortunately, in the increasingly competitive television marketplace, this has not happened.[4]

# History

Subscription TV has definitely had its ups and downs. After a rocky start, it experienced a boom and then lapsed into the doldrums.

The precursors of subscription TV have, like cable TV, been around almost as long as broadcasting. The form and shape of the medium has evolved over the years in such a way that the early experiments bear little resemblance to the present day model. In fact, some early forms of television channels for which viewers paid were wire rather than over-the-air so were a cross between what is now considered cable TV and subscription TV. The term subscription TV was not even common in early times. The more common terminologies were pay TV or toll TV. Along the way the medium also was dubbed feevee, a term invented by broadcasters to disparagingly differentiate the pay services from their own "free TV."

### Early Experiments

The first scrambling system called Phonevision was demonstrated by Zenith in 1947. In 1950 WOR-TV in New York briefly tested a scrambling system called Skiatron, and a year later KTLA in Los Angeles tested a system called Telemeter, which was owned by its parent company, Paramount.

The Zenith system, Phonevision, was given a ninety day trial run in Chicago in 1951 where three hundred families were hooked up by wire and given a choice of three feature films a day at a charge of one dollar per picture. The families did buy at the rate of an average of $1.73 a family per week, but the experiment was discontinued mainly because of technical problems. The Phonevision system involved using phone lines, and this turned out to be a cumbersome process through which the security could easily be broken, and people could see the films for free.

Paramount tried out its Telemeter system in Palm Springs, California, in 1953, but again the system proved faulty. Palm Springs subscribers had to place coins in a box in order to see the movies and also had a decoder box by the TV, which did not always work properly. About 40 percent of the subscribers viewed each film, so again the concept seemed to work but the equipment did not.

All of these experiments had set out to be just that—experiments. If they had succeeded, they probably would have been taken further, but they were not heavily promoted as money making ventures.

## Bartlesville and Etobicoke

However, in 1957 a high profile motion picture exhibitor, Henry Griffing, set up a pay TV operation in Bartlesville, Oklahoma, which was a profit-seeking, non-experimental nature. Griffing's system worked by connecting the subscribing homes by wires, but not wires associated with any cable TV system. Subscribers paid $9.50 per month no matter how much they watched. Griffing's plan was to have first run and rerun films, but unfortunately he was undercapitalized, which presented difficulties both in wiring Bartlesville and in securing films. His project failed and proved a major set-back for pay TV.

In 1960 Paramount decided to try an improved Telemeter system in Etobicoke, Canada. This time the project lasted several years, but the five thousand subscribers Telemeter was able to attract were not enough to ensure a steady profit.

## Regulation

Both motion picture exhibitors and broadcasters began to take note of these money making ventures, and because of the hoopla, particularly surrounding Bartlesville, they began lobbying for Congressional action or FCC action against pay TV. Motion picture exhibitors feared that people would not go out to movies anymore if they had them in the home, and broadcasters feared that pay TV would be able to bid more for the programming and talent available and thus destroy the programming of free TV.

The FCC decided it would take some action on pay TV and invited applications from commercial TV stations, which might want to test toll TV under certain circumstances. To appease the motion picture and broadcasting interests, the FCC stated that the tests were to run no longer than three years and were to be limited to markets in which there were at least four existing commercial TV services. No more than one pay TV operation was allowed in any one community so that pay TV would not become the rule rather than the exception. Also, the stations could not be entirely pay TV; they had to program at least twenty-eight hours a week of conventional unscrambled programming. To protect consumers in case the pay TV company failed or updated its equipment, pay systems had to lease decoding equipment to subscribers rather than selling it outright.

## Hartford, California, and Newark

The only applicant to come forth immediately, as a result of the FCC's invitation, was Zenith, coming back for another round with its Phonevision. This time it had a partner, RKO General, which acquired UHF station WHCT in Hartford,

Connecticut. Zenith and RKO began subscription TV programming in 1962 and continued it for the three year period. There were difficulties both in terms of equipment and programming, although the system did broadcast a few specials in addition to the movie fare.

An unfortunate page in pay TV's history occurred in 1964. Sylvester L. (Pat) Weaver, a former president of NBC, attempted a wired pay TV system in Los Angeles and San Francisco that would have brought movies and Dodgers and Giants baseball to its subscribers. It was in the beginning stages when the voters of California passed a referendum outlawing all subscription TV. This referendum was presented as a highly emotional, free TV versus fee TV, issue and was supported by theater owners who feared that pay TV would threaten their existence. Several years later the California Supreme Court declared the referendum unconstitutional, but the system had already been mortally wounded.

Adding insult to injury, in 1968 the FCC established more rules inhibiting pay TV including the rule that prevented pay systems from siphoning the broadcast programming by paying a higher price for it.

In 1974 Blonder Tongue Broadcasting tried to establish a subscription TV service on channel 68 in Newark, New Jersey, but it was never a success.

### The Peak of Success

Ironically, the turning point for subscription TV was pay cable, its present competitor in many areas. When Home Box Office succeeded in having some of the 1959 and 1968 FCC pay TV rules rescinded, including the antisiphoning rule and the rule that limited pay services to one per community, subscription TV received a shot in the arm. The general success of HBO also cast a glow on its cousin, now generally referred to as STV.

In 1977 ON-TV, owned by Oak Industries and Chartwell Communications, began operating on channel 52 serving Los Angeles with movies and sports. Another STV, SelecTV, was added to Los Angeles in 1978 and in rapid succession other cities such as Boston, Detroit, Phoenix, and Cincinnati eagerly adopted these pay services on UHF channels.[5]

ABC and its affiliated stations added an unusual twist to STV by receiving permission from the FCC to offer movies and other programming on a scrambled pay basis during the graveyard hours. ABC planned to make this programming available to its affiliates during the early morning hours when they are usually off the air. The intention was that subscribers would tape this material by setting the clocks on their videocassette recorders for the appropriate time. All in all during the late 1970s and early 1980s, STV looked like a successful phenomenon.

### Piracy

One of the problems to surface for STV in the early 1980s, however, was piracy. The decoder boxes which some of the companies supplied were not difficult to construct, and several enterprising engineers began manufacturing the boxes and selling them to the public.

This meant the people who purchased the boxes could view the subscription TV programming without paying the subscription TV company. Several of the companies took the box manufacturers to court and initially lost on the basis that STV signals are broadcast over the airwaves, which belong to all the people. On appeal, the subscription TV forces won and the box manufacturers were told to cease their practice.[6]

## Deregulation

In 1982 the FCC did an about face on its rules regarding subscription TV. In the interests of allowing STV to grow, it rescinded the rules that limited operations to communities with at least four commercial TV stations. This opened an additional 133 markets to STV that comprised 25 percent of the population.

The FCC also eliminated the rule that required STV stations to air at least twenty-eight hours a week of unscrambled programming. This means subscription TV stations no longer needed to form alliances with regular stations or in any other way concern themselves with conventional programming. They could, if they wished, establish twenty-four hour pay TV services.

The FCC also eliminated the rule requiring operators to lease decoding equipment and ruled that decoders should be sold by subscription TV companies to subscribers who wanted to purchase them rather than lease them. The reasoning behind this was that the sale would provide working capital for subscription operations and generally allow changes in business practices to meet marketplace demands. Technological developments regarding decoders had stabilized so the consumer did not need protection from equipment obsolescence.

STV operators approved of all the deregulation acts except the sale of decoders. They pointed out that this would lead to increased theft of their signals because piracy cases would be harder to prove and more difficult to keep track of if decoders could be purchased.[7]

## Decline and Fall

At first it appeared that this legislation and the general tenor of the media situation would be a huge boon for subscription TV. New stations were formed, some STVs began programming twenty-four hours a day, and ON began a satellite feed to service subscription stations throughout the nation.[8]

Soon hard times began to hit STV. The industry which had almost one million subscribers in 1983 had fallen to seven hundred thousand by 1984. No new systems were launched in 1983 or 1984 and many of the older systems closed, including several ON stations owned by Oak Media, the giant in the subscription TV business.[9]

Even the ABC venture, TeleFirst, designed to send programming into homes during the early morning hours, failed after a six month trial in 1984. During that time it never attracted more than three thousand subscribers in the Chicago area where it was tested.[10]

A number of elements were blamed for the fall of subscription TV. One was pay cable which became prevalent in the big cities as cable TV companies completed their wiring. A cable subscriber could receive all of basic cable and a pay channel for less than the cost of a single subscription TV channel. The popularity of videocassette machines was also cited as a factor because people could rent movies instead of paying to see them delivered over the air. Also, UHF independent stations became more profitable during this period. Station owners could make more money airing foreign programs, old reruns, or music videos than they could make with STV. The FCC ruling on decoder sales is cited as another factor, albeit less important. With people able to buy and own decoders, piracy was harder to detect.[11]

By the mid 1980s, subscription TV appeared to be a business on its way toward extinction.

## Issues

The main issue facing subscription TV is its very survival. An entity that appeared to be thriving only a few short years ago has hit upon hard times. Part of this is due to competition from other media and to its relative cost. In addition, its programming forms—movies, sexually-oriented material, local programming, and pay-per-view—have all experienced controversy. Signal pirates still cause economic hardship.

### Competition
Subscription TV's main competitors are the pay channels of cable TV. As cities become wired, more and more people switch from STV to cable because they can receive a multitude of channels from the cable service. Subscription TV used to overrun cable because it could become operational easier. By turning on a transmitter, it could cover a large area avoiding the expense and complications of stringing cable. But as cable overcame its wiring problems, STV became a less desirable commodity.

The wide penetration of videocassettes added to STV's miseries. For the consumer who only watches three or four movies a month, renting cassettes is much less expensive than subscribing to STV. Videocassettes also have the advantage of time convenience. The cassette can be viewed whenever the viewer wishes, not when the STV company schedules it.

Traditional broadcasting hurts STV also. Most cities which are large enough to support an STV operation also have many conventional TV stations. Although these stations do not air the same material as STV, they compete for audience. As UHF stations have been activated in large cities, they have further eroded STV's audience, and, in some instances, have abandoned the scrambled form of programming for advertising supported programming.

Added to this are even newer media, such as low-power TV (LPTV), direct broadcast service (DBS), and multipoint distribution service (MDS) which can offer the same type of programming as STV. Although none of these have been overly successful, they have, in some instances, taken away customers who might otherwise have subscribed to subscription TV.

Subscription TV seemed able to hold its own against the competition for a short period of time, but its future in the competitive ring seems very cloudy.

## Cost

Closely tied to the issue of competition is the issue of cost. Generally, a month's worth of subscription TV costs about twenty dollars. The main reason other entities have been able to take business away from subscription TV is that they offer more for less. A business that, by nature, can offer only one type of programming at a time, has a distinct disadvantage over a system which can offer multiple channels. A cable system's overhead changes very little whether it offers one channel or one hundred channels. The economics are such that subscription TV must charge almost as much for its one channel as cable charges for many.

In addition, broadcast TV is totally free to the public. Cassettes can be rented for as little as one dollar. These are economies of scale that a local operation like subscription TV can not match.

## Movie Depletion

Hollywood produces only several hundred movies a year, generally not even one for each day of the year. Many of these are less than sensational. On that limited supply, both STV and pay cable channels are trying to operate pay movie services up to twenty-four hours a day. The figures add up to disenchanted viewers who pay their twenty or so dollars a month only to find that frequently there is nothing worth viewing that they have not already seen.

Theoretically, more movies could be made, but whether or not more quality movies could be made is doubtful. There are only so many stories, so many truly creative writers to place these stories in script form, and so many creative actors and directors to make them come alive in believable fashion.

Movies, because they play first in theaters, must be geared to attract the young theater going audience and, as a result, may not be of interest to the older audience which stays home to watch TV.

Consumers generally do not understand the movie-making structure. They subscribe to a service which offers movies, fully expecting to be entertained with one scintillating movie after another. When this does not happen, they become disenchanted with the service and disconnect. Word of mouth negative publicity from these disconnects does not help the STV business.

## Sexually Explicit Material

Many STV's, partially in an attempt to stay afloat, have begun offering R-rated, X-rated, and unrated material. The arguments surrounding this type of programming are the same as those surrounding cable's presentation of this material. In fact, some programs developed for pay cable adult channels also play on some STV systems, and STV operators, like cable operators, have found that their subscribers are eager for this type of programming.

STV has limited such programming to the late evening, although some movies aired earlier in the evening have questionable content. Homes that do not want these programs can have their signals scrambled through addressability, but then they receive no STV programming during the hours of the sexually explicit programming. This has caused some discontent among STV subscribers because when an STV system changes to this type of programming, those not wishing it receive less programming than they would if the STV channel had stayed with movies and special events. On the other hand, subscribers who do want the adult material are often charged extra for it, a fact they resent.

On some systems this type of programming is the main money maker. The fact that the same programming is available on cassettes and cable, coupled with the controversy such programming usually engenders in a community, may indicate that STV is moving closer to extinction.

## Local Programming

At one time people felt that STV, because it covered a small area, could program worthwhile material of a local nature that would attract a large audience. That did happen in some cities with programming of both sports and concerts.

Other sports and music organizations were unwilling to sell rights to STVs, however, fearful that their gate intake might drop. As STVs hit upon hard times, they could no longer afford the rights fees demanded by the sports and events organizations, and the whole concept of local STV programming began to fall apart.

Some of the sports rights that once belonged to STV have been picked up by regional cable networks; other have gone to local broadcast stations; and in other instances games are simply no longer being telecast.

## Pay-Per-View

Pay-per-view has always presented logistical problems for STV. At first the hardware did not work properly, and connecting someone for a pay-per-view event was sometimes haphazard. Reliable addressability equipment solved that problem.

But, many STV customers resent the fact that they pay for an STV service with a limited number of interesting movies and then must pay an additional fee to see something that is above average. Other viewers would prefer an entire pay-per-view system, which would enable them to pay for only those programs they watched.

From the corporate point of view, pay-per-view is risky. Not only does it require extra promotion and staff, but the monthly income cannot be predicted as easily as it can with flat payments.

The pay-per-view concept has been around for a long time but has not been overly successful. Several prize fights have made a healthy profit for STV and other entities, but many times not enough people sign up to cover the added publicity and personnel costs.

One reason for the relative failure of pay-per-view may be the manner in which people have come to use the TV set as the universal pacifier. If people know they have to pay to watch something on TV, they will think before placing themselves before the set. This is contradictory to the way people usually use TV. Most people turn on the TV without thinking to amuse them, relax them, or help them pass their time. If all programming on TV were pay-per-view, TV viewing would decrease greatly, which could prove beneficial to society. But, as long as programming is available, which is free or which is paid for at a bulk rate, pay-per-view will have difficulty succeeding in any significant manner.

## Open Airwaves Concept

The fact that the STV signal is sent over the airwaves just like regular broadcasting hurts it both philosophically and economically. Many members of the public do not respect the signal as something privately owned which should be purchased.

A view prevails that the airwaves belong to the public, and, therefore, anyone has the right to view whatever is transmitted. Under such a philosophy, STV can not exist because it will have no economic base. Without advertising, its only source of income is subscriber fees. It cannot receive its programming for free because the movie production companies and athletic teams feel they have their due coming and will not offer their services or products for free. Unless STV charges subscribers, it can not operate.

But many subscribers do not feel that way. People renege on paying STV bills to a much greater degree than they do on utility bills. Somehow it does not seem nearly as criminal to steal a TV signal as to steal a commodity such as electricity or water.

Pirating activities were based largely on the open airwaves concept. Although selling illegal boxes is now a crime, the piracy issue still surfaces from time to time. Technically, there is no foolproof way to prevent piracy. The more complicated the scrambling systems become, the more ingenious the pirates must become, but that type of spiraling technology contest simply adds costs for the honest people who are paying to receive the programming. Although technology can aid in curbing piracy, the battle must really be fought on philosophical and legal grounds.

The scope of problems facing the subscription TV industry is staggering. But, except for a short period of time, that has been the history of this technology.

# low-power tv

# 7

## Description

Low-power (LPTV) stations are a new arrival in the television field. They are, as their name implies, regular TV stations broadcasting with very low power. They are still in the formative stage, but they have achieved great popularity with potential owners.

### Transmission Characteristics

Low-power TV stations broadcast on the regular TV channels, both VHF and UHF, so they can be received by anyone in their area who has a TV set. The VHF stations broadcast with ten watts of power and the UHFs with 1000 watts. Both types of stations can reach only fifteen to twenty-five miles in any direction, so they are definitely local in nature.

With this low power, they do not interfere easily with the high powered conventional stations, but the LPTV stations have secondary status, which means that if interference does occur, they must assume the responsibility by adjusting their power or going off the air. Obviously, the low-power stations cannot operate

**Figure 7.1.** LPTV configuration.

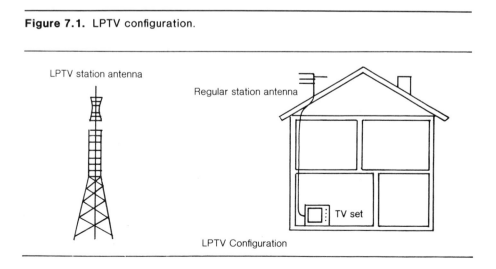

LPTV station antenna

Regular station antenna

TV set

LPTV Configuration

on channels already in use in their particular community, but they can broadcast on vacant channels provided they do not cause interference to stations that might be on those channels in other cities.

## Regulation

The FCC authorized LPTVs in 1980 with the intent of giving groups and individuals (particularly minorities) who have not traditionally been a part of the broadcasting scene a chance to participate.

The LPTVs are regulated by the FCC but with a loose hand. The FCC has made few regulations in terms of programming, finances, or ownership especially in comparison to the regulations imposed upon conventional stations. There are some programming regulations that LPTVs share with other stations, however. For example, LPTVs must adhere to the fairness doctrine by trying to give equal programming to all sides of controversial issues. They must also give equal opportunity and reasonable access to their airtime to all political candidates, but they need not supply production equipment to political candidates. They are also forbidden from broadcasting obscene and indecent material, they cannot conduct lotteries, and they are liable for copyright payments. Aside from those regulations, their programming can take just about any form.

They have no regulations concerning how they obtain their money. They can sell ads, operate on a subscription TV basis, ask for public donations, or gather money in other creative ways. There is no limit on the number of commercials they can show during their broadcast time.

No limit exists on the number of stations any particular group or individual can own. However, no one can own more than one LPTV in any particular area where their signals could overlap, and no broadcast stations or cable TV companies can own a LPTV in the same area where they have a station or franchise. The three TV networks are not to own LPTVs.

Overall, LPTVs are much less regulated than conventional TV stations. They do not need to talk with community leaders to ascertain the needs of the community as some full power TV stations do, and they do not need to supply the FCC with much of the paper work required of regular stations.

## Costs

LPTV stations are inexpensive to start and to operate. VHFs can be on the air for about $50,000 and UHFs for $80,000, much cheaper than the $2 million usually needed for a conventional TV start-up.

That cost does not include TV studio equipment, but LPTVs do not need to have studios if they don't want to. They can broadcast a schedule comprised totally of pre-recorded material. If they do buy studio equipment, they can use lower quality equipment than that needed by regular stations, equipment that is fairly low in cost.

The operating costs depend upon the type of programming the LPTV chooses. Stations that program religious material may be able to obtain totally free programming while stations that wish to create their own programs will find that their operating costs are fairly high.

## Programming

LPTVs have very few constraints covering what they program. In fact, they do not even need to program any specific number of hours.

They can obtain programming material from satellite services, from syndicators, or from their own studios. They can be as innovative or as conventional as they like.[1]

# History

Low-power TV has its roots in early television, but only since 1981 has it been a sought after commodity. Its future, although uncharted, could be promising.

## Translators

From the FCC's point of view, LPTV is an extension of translators—stand alone transmitters that have been rebroadcasting, mostly to rural areas, the programs of full-service stations almost since the beginning of television. These translators, some 4000 in number throughout the country, pick up the signal of a station in a nearby city, transfer that signal to a different channel and then send it out into the airwaves to areas of poor reception.

These translators had never been allowed to originate their own programming, but the FCC reconsidered and decided that existing translators as well as new facilities similar to the existing ones could engage in their own programming.

In 1980 the FCC issued guidelines for any translators that wished to begin originating programming and guidelines for people who wished to apply for new translators, now called low-power TV stations.

Groups wishing to apply for a new station needed to undertake an engineering study to find a place to locate their transmitter that would not interfere with conventional broadcasting. They also needed to establish a legal company, if one did not exist, which would be charged with the operation of the station. The FCC was particularly hopeful that women and minorities would become involved in this TV ownership so they would have a voice in their community. The FCC also wanted to know programming plans and, although it did not lay down any hard and fast rules, hoped that these would be designed to serve particular communities.

## Popularity

What happened was that a wide variety of groups and individuals applied for the stations not only women and minorities, but also large organizations such as Sears and Federal Express. Even though networks were not supposed to own LPTVs, NBC and ABC applied.

Some companies applied for over 100 low-power stations all over the country with the intent of forming entertainment networks, not at all what the FCC had originally intended. Others applied for large numbers of channels in order to rebroadcast VHF stations now on the air to other areas of the country. Many applicants stated that they planned to program their channels as subscription TV stations utilizing the programming of HBO or one of the other pay cable networks. Ted Turner applied in order to increase the coverage of his Atlanta super station and Cable News Network.

There were also many applications from women and minorities and from groups wishing to provide locally oriented public service. In Washington, D.C., eleven non-profit organizations joined together to apply for a station that would program public affairs, instructional, and cultural programming. A group of former FCC lawyers, all black, applied to form a network for minorities. A New York Chinatown resident wanted to offer Chinese programming on his station. The United Auto Workers planned to provide labor information, and a group led by Ralph Nader planned to program consumer information.[2]

In all the FCC received about 5000 applications by April of 1981 when it decided to impose a freeze on new applications so that it could sort out the applications already submitted. This was no easy chore because many of the applicants had engaged in "creative engineering" in order to prove that they would not interfere with conventional broadcasting, and the FCC felt the need to double-check the validity of these applications. In addition, many applications were for the same or similar areas so they would be interfering with each other.[3]

**Allocation of Channels**
After some study, the FCC decided it could allow construction to begin on properly engineered stations for which only one group or person had applied. As a result in December of 1981 the first low-power station went on the air in Bemidji, Minnesota, owned by a seventy-three year old retired broadcasting executive who programmed local news and sports and a pay movie service.[4]

The rest of the applications were studied further by the FCC which, in February of 1982 came up with guidelines concerning how the allocations would be determined. The Commission decided that any translators presently in operation could become low-power TV stations simply by notifying the Commission.

The FCC announced that the next applications to be decided would be those for rural areas, then those for small markets, then those for the large urban areas. The FCC stated that it would decide station applications by lottery but would give preference to applicants who did not own other media and to applicants with more than 50 percent minority ownership.[5]

The allocation of LPTV stations was conducted in an orderly manner and by 1985 some 80 stations were on the air in the lower forty-eight states and 194 were on the air in Alaska, many of them part of a network to keep people in the capital, Juneau, in touch with remote areas. Some of the continental U.S. stations were programming movies, but most were rich in local fare—dove hunting tips, Yugoslav-American customs, fishing advice, local news, high school sports. Most of the stations intend to sell ads and, although some have been successful, none are as yet profitable.[6]

## Issues

Virtually no one opposes the concept of low-power TV, but there are controversial issues, particularly in relationship to profitability, ownership, and competition.

### Profitability
Low-power stations, although inexpensive to establish and operate, must make money in order to survive. The most common method of doing this is through the sale of advertisements. But whether or not they can sell enough ads to make themselves profitable is certainly a reasonable question.

They will be competing with local newspapers, radio stations, and cable TV systems for the same advertising revenue. The other media will, in all likelihood, have been established earlier than the LPTV station so will have had a headstart in obtaining the advertising dollars. However, America is full of entrepreneurs who have managed to launch new ventures and, through hard work and enterprising activity, survive and prosper. Many of the new LPTV station owners may fall into that category.

Some LPTV stations are scrambling signals and offering pay services to subscribers. If pay TV is not already available in an area, this could prove to be a healthy business. If pay TV is already available, then the competition factor sets in again.

If LPTV stations are not able to become profitable, they will run the risk of being bought out by large companies who can build networks and can afford to keep a station operating without profit for a longer period of time. This leads to issues of ownership.

### Ownership
At present, the FCC seems to be holding to its original belief that LPTV should be for people who have previously not had access to media ownership.

The larger companies, however, are not content with this approach. The TV networks argue that they should be able to own LPTV stations just like anyone else. They have programming expertise which would enhance community programming. Other large companies point out their advantages in terms of assets to underwrite stations to insure their success.

If the small owners can not succeed with their stations, the bigger companies are standing ready. They would, in all likelihood, try to establish networks by buying out stations in a number of cities. This would greatly change the complexion of LPTV. Some feel that in order to prevent this, the FCC should institute rules limiting the number of stations which any one group can own.

### Competition

LPTV poses somewhat of a threat to older media such as radio and commercial television. It will be trying to obtain some of radio's advertising dollars, and it can be one more programming choice to draw viewers away from conventional broadcast stations.

It is, however, more likely to pose a competitive threat to cable TV. Both serve approximately the same size area in most instances, and both should be serving the local community. The type of programming cable TV local origination channels often cablecast is quite similar to what seems to be the dominant programming fare of LPTV stations. Both may be after the same audience and the same advertising dollars.

Overall, however, competing media do not consider LPTV as great a threat as some other media. Its low-power, local focus makes it desirable but not necessarily powerful.

# multichannel multipoint distribution service (MMDS)

## Description

Multichannel Multipoint Distribution Service is one of the smallest of the new television delivery systems in terms of number of viewers. It is sometimes referred to as "wireless cable" because the programming that it delivers is the same or similar to that of cable TV.

### Broadcast and Delivery

MMDS can deliver four channels of programming to buildings within a limited area. The signals are broadcast over the air in a manner similar to that of conventional television. The frequencies used for MMDS are between the 2000 and 2700 megahertz range—far above the conventional TV signals which range from 45 megahertz to 890 megahertz.

Because of this MMDS signals cannot be received on standard TV sets without some means of conversion. The means of conversion used for MMDS is called a down converter. The signals are sent out from an MMDS antenna at somewhere above 2000 megahertz, travel through the airwaves, and are received by an antenna geared to those same frequencies. Attached to the antenna is a down converter, which converts the high frequency signals down to the lower frequencies of a regular TV channel. In this way, the MMDS signals can be seen on channels 2 or 3 or 7 or 12 or any other regular channels. Usually, the channels selected for down conversion are ones not in use in the community to minimize interference.

### Distance

Because MMDS signals are at such high frequencies, they do not permeate as well as the regular VHF and UHF signals and reach only fifteen to twenty-five miles from their antenna. The reception point must be in line-of-sight with the transmitting antenna, which means that reception is poor if large buildings or hills are positioned between the sending and receiving antennas.

The frequencies presently allocated to MMDS can be used over and over in cities distant from each other because any one signal only travels twenty-five miles and will not interfere with a signal on the same frequency that is several hundred miles away from it.

**Figure 8.1.** MDS configuration.

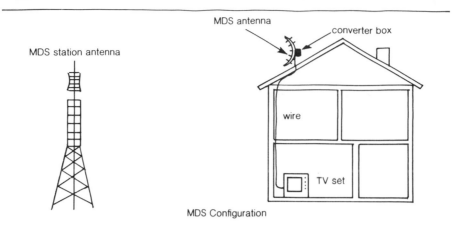

MDS Configuration

## OFS and ITFS

Two other technologies operate in the same frequency band as MMDS—OFS and ITFS. Operations Fixed Service (OFS) was designed by the FCC to serve businesses and similar organizations by enabling them to transmit primarily written data from one point to another. For example, a company might send memos from a company building located on the south side of town to another company building located on the north side of town. Three frequencies have been set aside for OFS but have not been greatly utilized.

Instructional Television Fixed Service (ITFS) is used for non-commercial educational purposes. For example, some school districts own ITFS systems and use them to transmit programming which teachers use in classes in school buildings throughout the school district. Medical colleges use ITFS to beam programs dealing with advancements in medicine to practicing doctors and nurses in hospitals. Universities use them to transmit course material to people at business locations who wish to upgrade their education. These viewers can watch the classes during their lunch hours or after work without having to commute to the university. Originally twenty-eight channels were set aside by the FCC for ITFS, and any single school system could own up to four of these channels. Like MMDS, these channels can be used over and over across the country because the ITFS signals only travel about twenty-five miles. ITFS was for many years underutilized which led MMDS operators to attempt to encroach on its territory.

## Programming

The programming which MMDS operators send out is usually the same as some of the programming which cable TV systems cablecast, primarily HBO, ESPN, CNN, C-SPAN, MTV or other cable satellite services. A MMDS system buys a satellite dish, makes contracts with several of the satellite services to use their programming, and transmits the programming from its antenna to antennas and down converters located on roofs around town.

For the most part, people pay to receive the MMDS channels, both by paying for the down conversion equipment installation and by paying a monthly fee for the four channels of programming. The initial installation costs have been approximately $250 and the monthly fee about $35.00. The MMDS company collects the fee and pays part of it to the program suppliers, just as cable TV companies do.

### Regulation

Because MMDS is an over-the-air service, it is regulated by the FCC, which decides, among other things, exactly what frequencies will be used for MMDS and who will be granted the licenses to operate the MMDS channels.[1]

## History

MMDS has had a rather complicated history in that it has changed form several times. Its structure is still somewhat in flux, and its progress is uncertain.

### MDS

MMDS originally started out as MDS (Multipoint Distribution Service). As such, it was to be a one-channel service rather than a four-channel service. This one-channel concept was established by the FCC in the 1960s when that body made some overall plans concerning the frequencies in the 2000 to 2700 range. It established the three services—OFS, ITFS, and MDS—the first for businesses, the second for non-commercial applications, and the third for commercial purposes.

Not much use was made of the MDS frequencies for many years, mainly because no one had ideas for making the channels profitable. With the advent of HBO and other pay TV services in the 1970s, several companies began to visualize a way of making money utilizing the MDS frequencies to transmit pay movies primarily to apartments and hotels. The down conversion equipment available at that time was so expensive that not many homeowners could afford to use it.

During the late 1970s, MDS systems were marketed on a limited basis primarily to apartment and hotel managers who then either charged individual tenants or guests who used the service or incorporated the costs into rent or room fees.

These MDS operations started primarily in areas that did not have cable so could not otherwise receive pay movies. The operation was sleepy at first, with one company, Microband, engaging in most of the MDS activity. But by 1980 business was brisk enough that the FCC began an inquiry into technical standards for MDS, methods for choosing from competing applicants for any particular channel, and possibilities for multi-channel MDS.[2]

## "Wireless Cable" Concept

The final point, which became known as multi-channel multipoint distribution service (MMDS), engendered the most interest. MDS operators saw that if they had enough channels to offer different services, they could act more like a cable TV channel with its multitude of channels. It was at this point that Microband began referring to the multi-channel concept as "wireless cable."[3]

In 1981, the FCC granted an experimental eight channel MMDS system to Channel View Incorporated of Salt Lake City. This test restricted Channel View to 135 receive locations and prohibited them from charging for their services. This test proved that an MMDS system could function, but it did not provide any information about the economic viability of such a system.

## ITFS Compromise

The main problem with developing a large number of MMDS systems was that there were not enough channels allocated to MDS to make it viable. However, many of the ITFS channels had not been utilized so several MDS companies and the FCC began eyeing the ITFS allocations as a potential source of channel growth for MDS. Educators did not take kindly to this idea and PBS acquired some of the unused ITFS frequencies in order to develop a network concept.[4] In 1983, however, after numerous compromises, eight of the ITFS channels were reallocated to MDS in cities where these eight channels were not already claimed by ITFS operators. ITFS systems were given permission to lease their excess time to commercial ventures. They were required to program three hours per weekday of educational material, but the rest of the time could be leased.[5]

## Lotteries

The FCC decided that two four-channel systems would be available in each market and that if there were multiple applicants for the systems, a lottery would be held to determine the award. The FCC did receive multiple applicants. In fact, some 16,000 applications were received from broadcasters, cable companies, telephone companies, and even a major league baseball team. Major markets such as New York and Los Angeles drew close to 150 applications, but even Anchorage, Alaska, attracted 66.[6]

Because of the large number of applicants, the FCC set up lottery guidelines early in 1985. The guidelines hold that no applicant can own both four-channel systems in any city, so two applications will be awarded in each city; applicants who win the lotteries can not immediately sell their property but must operate the MMDS for at least a year before selling; and minority applicants and applicants with no other media holdings will be given two-to-one preference.[7]

## MMDS-ITFS Systems

A few companies, however, are not waiting for the FCC lottery to be completed. They are establishing MMDS systems in cities by leasing time from established ITFS channels.

The first system to go on the air was in Washington, D.C. where American Family Theater, Inc. leased channels from George Mason University. Its four-channel line-up includes SelectTV, Cable News Network, C-Span, and the music video service Odyssey and Home Team Sports which share a channel.[8] Systems in New York, San Francisco, and Milwaukee which were one-channel operations offering HBO are making plans to go multi-channel.[9]

MMDS is still in an infant stage without a clearly defined market. While one MMDS operator feels MMDS can compete effectively with cable, another vows that it will only complement cable, broadcasting into areas that are inconvenient to wire. The whole area of signal piracy is only beginning to be explored.[10] The future will determine the course of this particular medium.


## Issues

Like many of the other new technologies, survival is an important issue facing MMDS. Also important are signal theft, the uneasy alliance with educators and with cable TV, and the lack of a clear-cut identity.

### Viability
The main issue facing MMDS is whether or not it can become a viable distribution source. In some ways it resembles subscription TV. Subscribers must buy special equipment to receive the signal. In the case of subscription TV, this is a descrambler box and in the case of MMDS it is a down converter, but the overall effect is similar. With subscription TV on the downslide, questions can be raised about the likelihood of another similar service building and surviving.

MMDS does offer four channels rather than the one offered by STV, but still not anywhere near the number of channels which can be offered by cable TV. The cost structure for MMDS is high at present and it probably will never be significantly lower than STV or cable TV. Just how many people will pay a $250 installation charge is unknown.

It does cost less to establish an MMDS system than a cable TV system because no expensive wiring is involved, but the cost of the down coverter equipment is still several hundred dollars.

MMDS may find its niche in rural areas which cable is not interested in wiring, but whether or not this is enough business to support an industry is questionable.

### Piracy
Theft of MMDS signals would be theoretically easier than theft of STV signals because the down converters are readily available for the ITFS market. MMDS operators are realizing that they need to scramble their signals for protection. This will, of course, add cost.

### Educational Alliance

An uneasy alliance exists between MMDS and the educational community. For starters, many in education are still a bit miffed that the FCC gave away some of the ITFS channels to this new upstart. Then, within the educational circles, there exists controversy as to whether or not educational institutions should lease their ITFS time to MMDS.

Educators are reluctant to be identified with brash commercial enterprises and worry that a channel which is owned by an educational institution may, during some of its hours, be programming unsavory material. This has already been an issue in Washington because the SelecTV service offered by American Family Theater programs some unrated films. These are a selling point for the MMDS operator but an embarrassment for George Mason University. On the other hand, leasing of time could prove a revenue source for educational institutions which are perennially underfunded. The time most desired by the MMDS operators is evening hours when many school activities are ended for the day.

### Cable Conflicts

Cable TV could design ways to make life uncomfortable for MMDS if the large cable industry decides that the small MMDS industry is a serious threat. So far MMDS has been careful with the cable issue portraying itself more as an ally than a foe.

### Visibility

MMDS faces acceptance problems with the general public. Anything with a complicated name like multichannel multipoint distribution service is not likely to become a household word. For this reason, some practitioners are now calling the service MCTV (multi-channel TV). The industry is small and can not afford expensive publicity and public relations. Consumers have been deluged with new media forms which have not kept their promises. Enterprising people with vision will be needed in this field if it is to succeed.

# satellite master antenna tv (SMATV)

## Description

Satellite Master Antenna TV (SMATV) is a small business that brings programming primarily to apartment complexes, condominiums, and hotels. Sometimes it is called "private cable" because of its similarities to cable TV.

### MATV
SMATV developed as an outgrowth of Master Antenna TV (MATV), a system
owners of apartment buildings had been using for years. In its simplest form,
MATV required a regular TV antenna installed on the roof of the apartment
building with wires from it to each apartment. In this way, all apartment dwellers
could receive the regular broadcast TV signals without each having to place an
antenna on the roof of the building or developing an elaborate rabbit ears configuration.

### Satellite Dish Addition
With the advent of cable TV, the apartment owner could buy a satellite dish to
bring in most of the cable TV network programming, hook this to the Master
Antenna TV system, and play some of the satellite programming on the empty
channels not occupied by broadcast TV. This combination of satellite signals and
broadcast signals became known as SMATV.

In a more complex form, owners of apartments, condominiums, mobile home
parks, hotels, and motels have build elaborate SMATV systems which wire together a whole series of buildings and are able to carry several hundred channels.
In this manner, a particular condominium complex can set up the equivalent of
a cable TV headend and become its own private cable system.

Usually an individual property owner will not handle such a complex set-up.
He or she will contract with a private company to provide the technical know-
how and service. This company takes on responsibilities similar to a cable TV
company except that it does not report to any governmental body. It contracts
with private individual property owners to wire private property.

**Figure 9.1.** SMATV configuration.

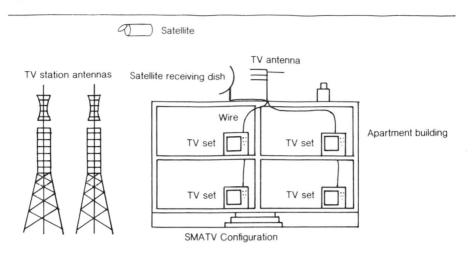

SMATV Configuration

## Regulation
Although SMATV companies compete with cable companies for business from the same buildings, SMATV does not come under the FCC definition of a cable TV system. The FCC's definition specifically excludes systems that serve fewer than fifty subscribers or serve subscribers in one or more multiple unit dwellings under common ownership, control, or management.

This means SMATV operations are unregulated. They are not required to pay fees to city governments and they are not required to establish any access channels. In practice, they sometimes pay a percentage of their income to property owners in exchange for the right to build and operate the SMATV system, but this is less expensive than the franchising fees cable TV systems bear. Some of them have also established access channels which program features about the residents of the establishment or stories about local services or events.

## Economics
In theory, at least, SMATV systems can be economically sound. If an existing MATV system wants to add five satellite services to its existing broadcast stations, it can do so for a cost of about $35,000, most of which is spent for purchase of the dish. If a hundred customers pay $25 a month, the SMATV will recoup $30,000, or almost its total investment, in only one year. Even if an apartment complex is wired from scratch, the investment can usually be recovered in a few years.

## Programming
The SMATV system uses the same programming as the cable TV systems. A system adding on to a MATV will operate the same as a twelve channel cable

TV and simply cherry pick from the satellite services, perhaps offering a pay movie channel, ESPN, CNN, and Nickelodeon. Newer systems will include a much larger selection of programming. SMATVs are more likely than cable systems to include adult channels because they do not have the watchful eye of government looking over their shoulder. Some SMATV systems even have two-way addressability which they use primarily to add or subtract customers.

At present, less than a million households receive television through an SMATV system. This is a business which is growing slowly but steadily.[1]


## History

Satellite Master Antenna TV has had a short and rather unauspicious history. It began quietly and, except for a few conflicts, has proceeded quietly.

### Early Roots
Satellite Master Antenna TV became possible in 1979 when the FCC deregulated receive only earth dishes by stating that no one needed to apply for a license in order to install and use an earth station which received satellite TV signals.

Exactly who was the first to buy a dish and connect it to an apartment building's master antenna TV system is unknown, but SMATV grew slowly but surely to a subscribership of close to a million by 1985.[2]

Many SMATVs were begun because cable was not imminent in the area, and others were begun because apartment owners were wary of cable TV companies defacing their property with unsightly wires and holes in walls. Generally SMATV operations were able to achieve 50% penetration, if the system was addressable. If not, the penetration was lower because renters stole the signal by tapping into a neighbor's hook-up or the general wiring system.

### Original Builds
Most SMATVs were small, unglamorous set-ups added on to already existing MATV systems or built into new modest-sized apartment complexes. In 1982, however, the bid was let for a very large SMATV system called Co-op City in New York's Bronx area. It consisted of 35 buildings, 15,000 apartments, and 60,000 people. The company which won the bid, Satellite Television of New York, outbid six other companies and did agree to pay Co-op City a percentage of its income.[3] Co-op City is still the largest SMATV system, but other fairly large ones have sprung up in Dallas, Atlanta, and other cities.

### Cable Competition
Throughout its short history, SMATV's primary antagonist has been cable TV. The two compete in attempts to wire large complexes such as apartments and condominiums. Cable has resented the fact that SMATV is unregulated and has made several attempts to include SMATV under both FCC and city government

control. In 1983 a cable TV system did succeed in getting a New Jersey superior court judge to rule that an SMATV operator could not operate without a cable permit. That ruling would have made SMATVs subject to the same local government regulations as cable TV. However, in 1984, the U.S. Court of Appeals reversed the decision exempting SMATV from state and local regulation.[4]

Cable satellite services have also impeded SMATV's growth by refusing to sell them programming. This was particularly true of the movie services—HBO, Showtime, and The Movie Channel. SMATV felt discriminated against and pointed out that the parent companies of the movie services—Time Inc., Viacom, and Warners respectively—had large cable TV system holdings which needed protection from SMATV. The satellite services replied that their cable systems had nothing to do with withholding programming and that the issue revolved around the fact that SMATVs were too small to bother with and sometimes did not pay their bills. Gradually the dam broke. Shortly after Showtime and The Movie Channel merged, they began an active pursuit of the SMATV market mainly because it represented added revenue. In 1985 HBO loosened its restrictions and began courting the SMATV market also.[5]

Another bone of contention between SMATV and cable has been access to apartment complexes. Both have wanted exclusive, inexpensive access. Here the courts and legislative bodies have generally sided with SMATV. In 1982 the Supreme Court ruled that cable TV systems could be required to pay apartment owners just compensation for access to their buildings to wire them with cable. This "just compensation" has been left to state courts to determine, but generally it has made the wiring of apartments more cumbersome and expensive for cable TV companies.[6]

**Congressional Action**

In 1984 Congress was even kinder to SMATV. The cable TV deregulation bill originally contained a clause which prohibited SMATVs from having exclusive agreements with landlords. Through lobbying efforts by SMATV forces and the real estate industry, this clause was struck. If SMATV operators do make agreements with landlords giving them exclusive rights to wire a complex, the door will be further closed for cable's penetration into the markets that SMATV desires.[7]

The kind of treatment that SMATV has received is probably due in part to its underdog status. Because its only market is multi-unit dwellings, it will not become a giant business. However, bankers are now willing to loan SMATV companies money and the industry has achieved a generally respectable status. Cable TV operations continue to eye it with suspicion and a touch of fear.

## Issues

The main issues facing SMATV involve its adversarial relationship with cable TV, although piracy and survival are also concerns.

## Cable Conflicts

As cable sees it, SMATV is capable of taking away large portions of cable's income by signing exclusive contracts with owners of apartments and other complexes. These owners would then be able to prevent cable TV from entering the premises. Cable claims it needs the apartments and condominiums in order to recoup its costs, especially in the big cities where it has made extravagant franchising promises.

SMATV retorts by saying that apartments had always been lowest on cable TV's priority list, and that cable often wires apartments after all homes are wired because it doesn't like the high turnover rate, and hence high disconnect rate, of apartments. Now that someone else is expressing interest in apartments, cable is acting like a scorned suitor. In addition, the SMATVs need the exclusive contracts so that they can survive economically. SMATV argues that if cable is allowed to offer service after an SMATV system has been established, the cable system can cut rates temporarily, while depending upon its well-heeled owners to absorb the cost. The customers will switch to the lower-priced cable, forcing the SMATV out of business. Then the cable company will raise its costs, and both consumers and SMATVs will lose.

Cable is particularly bitter about the regulation it must endure when SMATV has none. Cable interests point out to cities that they should make attempts to keep SMATV from growing because loss of revenue to cable TV companies is loss of revenue to cities through lower franchise fees. If cable TV had its way, cities would impose an "entertainment tax" on SMATV. It also cries that SMATV, left unregulated, is likely to build systems which are not technically sound or to program undesirable material.

SMATV, of course, rides on the free enterprise coattails pointing out that it does not need regulation because landlords are free to choose whether or not they want SMATV and further, are free to choose to allow wiring by cable TV companies.

## Piracy

A less serious problem for SMATV is theft of signal. One apartment dweller can fairly easily tap a signal from a neighboring apartment if the equipment used by the SMATV system is even slightly unsophisticated. To counter this, most SMATV systems use addressable equipment to control and detect which apartments are receiving the TV signals.

## Survival

The SMATV business does have to be concerned about overall survival. It is a small industry without a great deal of organizational support. A change in legislative or judicial mood could easily impact upon its long term health.

# direct broadcast satellite

# 10

## Description

Direct Broadcast Satellite (DBS), unlike many of the other new technologies, has a very high entry fee. It involves expensive high-powered satellites intended to beam directly to the consumer. This technology is not operational at present, but it is on the drawing board at a number of companies.

### Transmission and Reception
DBS is a service that plans to transmit directly from satellite to home receivers. At present some people have placed satellite dishes in their back yards or on their rooftops in order to receive satellite programming, and have thus set up their own DBS. But the programming presently delivered by satellite is not intended to be received directly in the home. It consists primarily of programming meant for cable TV as well as broadcast network feeds and special feeds intended for other companies. Some of the other newer media such as MMDS and SMATV are also authorized users of the satellite feeds.

DBS would be a service intended for the end user—not an intermediary such as cable TV. The programming would be beamed up to a satellite transponder from an earth station in the same way that cable satellite programming is now sent. However, the satellites to which it would be beamed would be much larger and more powerful than the present cable-oriented satellites. Satellites create their own power from the sun, so in order to increase power, the size of the satellite must be increased to catch and convert the sun's rays. The DBS satellites are planned to have only three transponders rather than the present twenty-four, but each transponder will be much more powerful than the present satellite transponders.

The satellites will also transmit in a different frequency range from the present cable-TV oriented satellites. DBS is planned for the 12 gigahertz range which is in what is referred to as the Ku-band. This is higher up in the electromagnetic spectrum than the C-band at four to six gigahertz which the cable satellites use.

The DBS transponders will send the signals back to earth to rooftop receiving dishes which are about two and one-half feet in diameter. This is in contrast to the ten-foot dishes which are now used for satellite reception. This smaller sized dish is possible because the DBS satellites will be more powerful than the present

**Figure 10.1.** A DBS reception process.

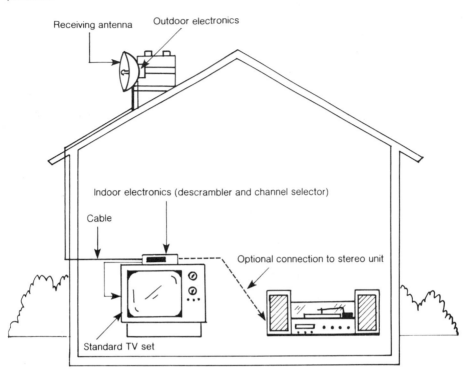

Receiving antenna

Outdoor electronics

Indoor electronics (descrambler and channel selector)

Cable

Optional connection to stereo unit

Standard TV set

satellites and because dishes receiving Ku-band can be smaller than those used for C-band.

On the back of the receiving dish will be a unit that will downconvert the frequency of the satellite signal to a regular TV channel frequency and send this signal by wire to a TV set in the home. If the signal has been scrambled, a unit attached to the TV set will unscramble it. All of this, of course, circumvents the present intermediaries between programming source and consumer, most notably the broadcast stations and the cable TV systems.[1]

### Finances

A huge start up investment is needed to enter the DBS field. A great deal of engineering goes into the design of the satellites and the receivers. The satellites must be built, launched, and fine-tuned. Programming needs to be procured, and receivers manufactured. The companies entering the field figure that at least $700 million will be required to make the service operational, and that $700 million will cover only one satellite beaming to one time zone. Four satellites will be needed if programming is to be geared to clock time of each zone.[2]

## Programming

The proposals which have been made for programming include the old standby, pay movies, as well as children's programs, sports, education, culture, and general entertainment.

Some of the companies wishing to become involved in DBS are planning to act as common carriers and lease their transponders to companies wishing to provide programming. Other companies are planning to set up their own full-scale networks, some on a pay basis and some supported by advertising.[3]

## Regulation

The FCC is the governing body which will oversee DBS. It has already issued some tentative rules which state that DBS licenses will be issued for periods of five years and that anyone receiving a license must be actually operating within six years from the time the license is first granted. There are no limits on who can or can not own a DBS system and no limits on how many DBS channels any one entity can own. DBS will be subject to the fairness doctrine and the rules governing the appearances of political candidates, but overall regulations governing DBS will not be oppressive.[4]

# History

DBS does not have much of a history because it is not yet operational. However, it has been discussed and planned for a number of years, and has had its share of ups and downs.

## STC Proposal

Rumblings toward DBS date back to 1979 when the FCC deregulated receive only earth dishes stating that no one needed to acquire a license in order to install a dish to receive satellite signals. When that occurred, a company called Satellite Television Corporation owned by Comsat, the satellite launching organization, began discussing the establishment of a programming service to beam signals directly from satellite to homes.

The three networks rose up in arms over the idea, primarily because they saw this as direct competition. They predicted the downfall of localism because DBS systems obviously could not program local material. If DBS caused the demolition of the network-affiliated stations structure by programming directly to homes, no one would be engaged in local programming.[5]

In 1980, Satellite Television Corporation, undaunted by the network criticism, issued a formal proposal to the FCC asking for permission to develop a programming service which would go directly to homes starting in 1985. This service would be a pay service using scrambled signals. Three channels of programming would be available. One, the Superstar channel, would be similar to existing pay cable channels featuring movies, concerts, and specials. A Spectrum channel would include film classics, public affairs, children's programs, performing arts, and other cultural fare. The third channel, Viewer's Choice, would feature sports, adult education, and experimental theater.[6]

The broadcasters voiced opposition once again, and called attention to the fact that the international Region 2 Administrative Radio Conference was scheduled for 1983 to discuss international allocations of satellite space and that decisions regarding satellite broadcasting should be postponed until after that conference. But the FCC disagreed feeling that the United States should have a firm idea of what it was planning to do when it attended the Conference. As a result, the FCC approved STC's idea and invited others to apply to enter the DBS arena also. And who should apply but CBS, RCA, and other names familiar in the broadcast world.

## Other Proposals
In all, the FCC received thirteen applications and approved eight of them. The ones they did not approve were rejected primarily because they did not contain adequate information.

Of the eight proposals accepted (nine counting the previously approved STC proposal), three of them—Western Union, RCA, and Direct Broadcast Satellite Corporation—planned to own the satellites but lease the transponders to others for programming purposes.

U.S. Satellite Broadcasting Company, owned by the Minnesota-based broadcaster, Hubbard Broadcasting, planned to begin a fourth network that would be intended not only for homes but also for affiliated stations. It would be advertiser based and the broadcasters who received programming from it would also produce programming for it.

Dominion Video Satellite Systems, Inc. and Graphics Scanning each proposed two channels of commercial programming, primarily of an entertainment and sports nature.

Focus Broadcasting planned to first begin a direct broadcast service on a Westar satellite in the early 1980s and then transfer this service to a DBS satellite when it became available in 1985. The planned service would consist partially of pay movies and partially of advertising supported general programming.[7]

## High Definition Proposal
The most controversial of all the proposals was the one from CBS. It suggested that all direct broadcast satellites should beam high definition TV signals which would provide pictures of over 1000 lines of resolution instead of the present 525 lines. CBS's argument for this was that high definition TV technology exists and should be implemented, and here was the perfect new technology to devote to this improvement.

The other DBS applicants did not agree with CBS, primarily because high definition TV uses much more spectrum space and would, therefore, greatly reduce the number of channels available in the space provided for DBS. They also pointed out that regular TV sets would not be able to receive the high definition signals so DBS would be dependent upon people buying new sets.

NBC and ABC came to CBS's defense, as did some of the broadcasting organizations, saying that programming should be placed simultaneously on DBS and on conventional networks allowing the turnover to high definition sets to take place gradually. The FCC rejected CBS's idea that all DBS should be high definition but did grant CBS permission to experiment in this area.[8]

## Interim Approval

The interim approval for all these projects was given by the FCC in June of 1982, dependent upon the outcome of the 1983 Administrative Radio Conference. This conference proved satisfactory to the United States and it received enough satellite space to proceed with DBS.[9]

The companies which had been given the go-ahead began making their plans to be in operation about 1985. The FCC mandated that a "due diligence" report be given in 1984 to make sure that the companies were progressing toward DBS service and not simply warehousing the frequency space allocated to them.

## United Satellite's Operation

Into this plan came a maverick—United Satellite Communications, Inc. In 1982 it petitioned the FCC to begin a rather vague type of TV service which the FCC approved. Soon it had purchased time on a Canadian mid-power satellite and set up a DBS system.[10] The Canadian satellite required only a four foot receiving dish as opposed to the ten foot dish required for the American satellites delivering to cable TV systems.

United set up five channels of service—two movie channels, ESPN, and two news and information channels and in the fall of 1983 began marketing them in Indiana. They charged customers $40 per month for the programming plus a one-time rooftop dish installation cost of $300.[11]

Comsat and others plodding their way toward the high powered DBS systems objected, but to no avail. United moved their marketing into other parts of the country, but the results were weak at best and six months after its launch it had only 1000 customers.[12] After struggling for a year and a half, United went out of business, leaving the few thousand subscribers it had attained with dishes but no programming.[13]

## Due Diligence Tests

Meanwhile, several of the original DBS applicants flunked their "due diligence" test, mainly because they had not contracted to have the high power satellites built. Others did not even bother to fill out the FCC papers because they were no longer interested in DBS.

When the 1984 deadline passed, only Comsat's Satellite Television Corporation, Direct Broadcast Satellite Corporation, Hubbard's U.S. Satellite Broadcasting Company, and Dominion Video Satellite Systems were still in the FCC's good graces and given permission to proceed.[14] CBS, which had been eliminated, flirted with the idea of joining forces with Comsat and its Satellite Television Corporation but decided against the venture.[15]

Then, in a very surprising move, Comsat, the original pioneer in DBS, announced it was pulling out of the business after having spent five years and $140 million gearing up for it.[16]

This, combined with United's demise, made the DBS industry look very shaky. Although the companies still interested in DBS affirmed their belief in the medium, the mid-80s found the whole issue of satellite to home reception very muddied.

## Issues

The main issue involved with DBS is the shaky potential for its implementation. If it does become successful, then its relationship to cable TV and to broadcast TV will become crucial as will as its attempts to obtain programming.

### Implementation
Given the recent events in the DBS arena, the success of the medium seems far from assured. The enormous investment needed to initiate the service would make the most well-heeled of companies think twice. The revenue to pay back the investment must come from subscribers willing to buy dishes and pay for the programming services or from advertisers. Most of the potential subscribers already have many programming options open to them. Convincing them to switch to DBS will take aggressive marketing. In a similar manner, advertisers seem happy with broadcasting and other media and may not be willing to experiment with a new, untried form.

The potential technical problems associated with DBS also could cause long term and short term delays. The 12 gigahertz range is not well developed, so technical problems such as interference could occur. Undoubtedly, engineers can design around these problems, but this debugging will waste precious time for a medium that is already a johnny-come-lately. Windstorms, which tear down broadcast antennas, can also tear down satellite dishes. Climatic conditions and breakdowns on both the sending and receiving ends may make DBS unreliable.

### Relationship to Cable TV
The prime competitor with DBS seems to be cable TV. Both plan to offer satellite programming services of a similar nature. Cable systems are building rapidly, at least in comparison to DBS's progress. There is a question as to whether or not people who just took down their broadcast antennas to bring wires into their homes will now discard the wires and put up another antenna. Cable TV interests think they will not and should not.

To counter this argument, DBS has, at times, pictured itself as a competitor with subscription TV and MMDS rather than cable. It sees itself as a service to rural areas which are not worth the cost of cabling. If this, indeed, becomes DBS's target audience, it may be carving itself a market so small that it will never cover the initial start-up costs.

## Localism
If DBS becomes successful, it has the potential of replacing the networks and, therefore, the local programming of their affiliated stations. This could lead to a demise of localism within the media. The main programming fare would be coming from a national source without opportunity for input and discussion concerning local issues.

DBS has refuted this localism issue by pointing out that many entities are engaged in local programming now, including not only broadcast stations but cable local origination and access channels and low-power TV stations. They claim localism has nothing to fear.

## Programming
For the most part, the DBS programming proposals mirror those of broadcast and cable TV. DBS could be one more source competing for the same Hollywood movies everyone else programs.

The one unique DBS proposal was rejected by the FCC, but may be able to stage a revival given the present lethargic state of DBS. CBS's plan for high definition TV did have some merit. It would have allowed high definition TV to develop alongside regular TV and would have provided a natural turnover to TV sets that could render much higher quality pictures. If the programming placed on the DBS satellites were programming already available (say, perhaps, programming from NBC, CBS, and ABC), then people with old sets could watch in the traditional way, and people with new sets could buy satellite dishes and watch the high definition version of the programming. Eventually everyone would convert to high definition sets, and the technical quality of television would improve. This type of arrangement would take more spectrum space and therefore reduce the number of channels available for DBS, but given the fallout that has occurred in the industry, all the channel space may not be needed.

DBS must solve its technical, financial, and programming problems in order to become a competitive force. Whether or not it can do this remains to be seen.

# videocassettes 11

## Description

Videocassette recorders were an outgrowth of videotape recorder technology so have been utilized in professional and consumer configurations. This discussion will center primarily on the consumer market.

### The Video Taping Process

Videotape recorders stem from audio tape recorders. Audio tape recorders are devices that rearrange the iron particles on magnetic tape so that sound impulses can be stored on the tape and played back at a later date. This rearrangement of particles is undertaken by stationary heads, ones that record new material and ones that erase old material. Audio information is fairly simple, so that quarter-inch tape moving at the rate of seven and a half inches per second past a stationary head can have its iron particles adequately rearranged to represent the audio material.

A video signal, however, is quite complicated and contains much more information. If video were to be recorded by stationary heads, the tape would have to move at fifty-five feet per second, or nearly thirty-six miles per hour. A one-hour videotaped program would require approximately 198,000 feet of tape, obviously, an awkward amount to deal with. To overcome this problem, engineers designed methods of moving the heads so that information could be placed on the tape diagonally or vertically instead of horizontally, as is the case with audio tape.

The first method developed placed information vertically on two-inch tape. Four rotating heads were used, each picking up where the other left off, so that impulses were applied to the tape constantly. This method was known as transverse quadraplex and was used in broadcast facilities for many years.

The second method developed was called helical and placed material on the tape diagonally. For this method tape from a supply reel was wrapped around a drum at a slant and then run onto a take-up reel. The drum contained a number of heads, which spun at a rapid speed and placed the video impulses on the tape at an angle from top to bottom. Audio information was placed on the tape by stationary heads, and information contained on a part of the tape called a control track assured that the picture remained stable when it was recorded and played back.

The first helical recorders used one-inch tape, but as the technology improved, they were able to operate effectively with three-quarter-inch tape and then half-inch and quater-inch tape. Sometimes this tape was in an open reel configuration, meaning that the tape operator had to thread the tape from a supply reel past all the appropriate heads to the take-up reel. Other times the tape was in a cassette, which is a closed container, that does not need to be threaded by the operator. When the cassette is placed into the videocassette tape recorder, it threads automatically by the use of various levers, which pull the tape from the cassette and thread it around the appropriate heads.

Most of the tapes that are used for the consumer market are half-inch cassette tapes that are placed in videocassette recorders (often called VCRs) which use the helical principle.

## Formats

There are two different half-inch formats of home videocassette recorders, and, unfortunately, the two are not compatible with each other. Tapes recorded on one type of machine will not play on the other.

The two formats are called Beta and VHS, and they differ primarily in how the tape threads. Beta tape decks have a single swinging arm that draws the tape from the cassette and wraps it around the drum. VHS tape decks take the tape out with two parallel arms, which then hold that tape to the head drum in what is called an "M-load" configuration because it resembles the letter M. Both of these loading patterns have their advantages and disadvantages. The Beta system takes longer to thread initially, but can go instantly from play into rewind or fast forward. The VHS pattern threads quickly but must unthread before it can go into fast forward or rewind, and then rethread again before going into play.

The VHS recorders have sold at about three times the rate of the Betas making software for them easier to find.

## Features

The videocassette recorders developed for the home market contain a wide variety of features. The most basic of recorders will record, play back, fast forward, rewind, and stop, all accomplished with very clearly marked buttons. Beyond that an almost infinite number of options exist. Of course, the more functions a recorder performs, the higher its cost.

One of the most popular features is a clock, which enables the owner to set the cassette recorder to tape at a particular time. This means that the VCR can tape a program broadcast over-the-air while the owner is not at home. This taping for later viewing has become a primary function for most VCRs. Some VCRs can be programmed to tape a number of programs on different days, at different times, and on different channels. The owner can leave for a week or two and come back to a number of hours of taped material.

Most Beta and VHS decks also offer the user a choice of recording speeds so that between one and six hours can be placed on a single tape. The tape is the

**Figure 11.1.** Beta and VHS tape
threading patterns.

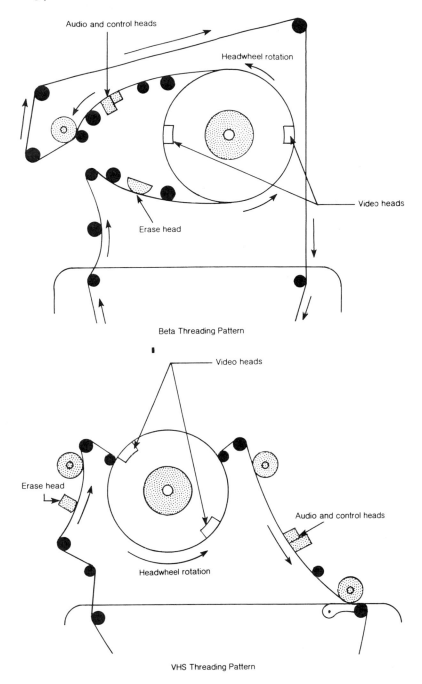

Audio and control heads

Headwheel rotation

Video heads

Erase head

Beta Threading Pattern

Video heads

Erase head

Audio and control heads

Headwheel rotation

VHS Threading Pattern

same, but the speed at which it moves through the recorder differs. The slower speeds use less tape per moment of recording but sacrifice some of the picture quality, while the higher speeds attain better picture quality but at the expense of more tape.

Other features common on recorders are still-frame, frame-by-frame advance, slow motion, and fast motion. Still-frame enables the user to hold one particular frame on the screen for examination. If the frame is held too long, the tape will become overly worn, so most recorders automatically turn off the still-frame after several minutes have elapsed. Frame-by-frame advance allows the tape to be viewed one frame at a time so that particular motions or events can be found. Slow motion is accomplished by playing back at a slower speed than was used when the material was recorded.

Fast motion, also called high-speed scanning, is a frequently used feature. It enables the operator to proceed through a tape either forward or backward and still see the images. Obviously, the images are moving more rapidly than in the play mode, but this high-speed scanning feature allows someone to find a particular spot in a program easily and quickly. On some machines the speed of the high-speed scanning can be varied so that the image moves only slightly faster than normal to a speed that may be five times normal. In connection with this, some machines play the audio as well as the video at a rapid speed. If the speed is not too great, the audio can be understood, and the material can be comprehended in less time than it would take if the program were watched during its regular broadcast.

This high-speed scanning can also be used to fast forward through commercials when playing back a tape. This practice is not in favor with advertisers. Other recorders have a pause button, which can eliminate commercials during recording, providing an operator is present at the time of the recording. When the commercials begin, the operator places the recorder in pause, and all recording stops, although the tape remains threaded so that recording can begin again at any moment. When the commercials are over, the operator pushes play and the program continues, commercial free.

Many VCRs can record one program while another is being viewed because the VCR has a built-in tuner, which is separate from the tuner of the TV set. Therefore, the VCR tuner can be set for one channel and the TV set tuner for another.

Some of the newer VCRs have a higher quality stereo sound than that achieved by audio recorders. This is because they playback the sound, as well as the pictures on the tape, diagonally. This means more room is used for sound and, therefore, it can be better quality. Of course, the sound has to first be recorded diagonally, but more and more preproduced cassettes are being recorded in stereo. They are also recorded in the traditional monarual method so they can be played back on older VCRs.

All VCRs contain tape counters that help the users locate particular places on the tape, providing they noted the numbers at the time of recording and zeroed the counter properly. In addition, some cassette recorders place a silent tone on tapes at places where particular programming begins or ends. The tape recorder can then be placed in fast forward and will stop at the point of the tone. For example, if three programs are recorded on one tape, a tone could be placed at the beginning of each program. The tape recorder then could fast forward to exactly where each program began.

Many of these functions can be operated through remote control devices attached to the VCRs.

Cameras and microphones can also be connected to VCRs allowing owners to tape their own material. About one VCR owner out of four buys this additional equipment. Microphones come in all gradations of quality, and cameras offer a wide choice of options—zoom lenses, viewfinders, filters, and battery packs. In some instances a camera and a recorder are one unit (camcorder) that is compact and lightweight. Some recorders have an audio dub position so that voice-overs can be added to taped material after the fact. Two recorders can be connected together for editing with various levels of sophistication built into the editing controls.

Anyone in the market for a videocassette recorder should think carefully about the uses to which the recorder will be put so that the proper features can be purchased. The cost of a videocassette system can run from several hundred dollars to several thousand dollars depending on the options selected.[1]

## Uses

The primary reason most people use their videocassette recorders is to tape programs off the air while they are not at home. Some of these programs are viewed later; others fill a closet of good intentions but are never watched.

Other uses for VCRs include recording one program while watching another, recording programs while they are being watched, making "home videotape" recordings, and playing pre-recorded material rented or purchased from video stores.

The home taping market has been particularly attractive. Instead of using super 8 cameras for home movies, people are using video cameras and their VCRs. The results can be seen instantly—no film to be developed, and the tape can be erased and reused at a later time.

Pre-recorded software for videocassette recorders has grown at an increasing pace. The availability of this material was not crucial to the initial sale of videocassette recorders because most people were using the recorders for off-air tapings. However, as the quality of software improved, the quantity of its sales increased.

The pre-produced programming that does exist is predominantly movies—the same movies that appear on pay cable and subscription TV. In addition, sexually explicit movies have proven to be very popular. Self-help and individual interest materials are also available as are educational materials. Several unique applications of videocassettes have arisen such as a dating service, which circulates videocassettes in which people talk about themselves and the type of people they would like to date.

Pretaped programming can be purchased or rented at many audio stores, which have added videotapes to their line, and at video stores, which have sprung up to serve this new market. In addition, some libraries loan videotaped movies. The rental price of movies falls around one to five dollars while the purchase price might range from twenty to one hundred dollars. Blank half-inch cassette tapes sell for about seven dollars.

In 1984, Americans bought about 100 million blank videocassettes, and seventy-eight percent of VCR owners either rented or purchased prerecorded programs on cassettes.[2]

### Regulation

Videocassette recorders are not regulated by any government bodies such as the FCC or local city councils. The price of both machines and tapes is dependent upon cost, competition, and what the market will bear.

The cassette market has experienced government attention in the form of a copyright court case and related Congressional hearing, but formal regulation is not part of the industry.

The characteristics of videocassette recorders and the videocassette industry are fairly well set, but innovations will no doubt continue on a limited scale.

## History

Home videocassette recorders have had a fairly easy time finding their niche among the new technologies. Because they were an outgrowth of videotape recorders, they had a chance to test their wings in professional broadcast TV, education, and industry before being introduced into the consumer market.

### Early Recorders

The first videotape recorders introduced by Ampex in 1956 used two-inch tape and were very expensive, bulky, and difficult to operate. The tape also had to be threaded from one reel to another past various audio and video heads, a rather tedious process. In the late 1960s and early 1970s several companies introduced videotape recorders which used one-inch, three-fourths-inch, or half-inch tape. Some of these had to be threaded, and others were of a cassette nature and threaded themselves after the operator inserted the tape into the machine. These machines were used primarily by educational institutions and industrial companies, for although they were cheaper and simpler to operate than the two-inch machines, they still needed an engineer handy for their care and feeding.

Then in November of 1975 Sony introduced its Betamax machine, a half-inch videocassette machine specifically designed for the home consumer market. The intial Betamax'used tapes that could record for one hour and sold for $1300, but the price was actually $2300 because the recorder could only be purchased with a new Sony color TV set.[3]

## Competition

The price went down and the features went up as Sony received competition from Matsushita, which in 1976 introduced a cassette recorder it called video home system (VHS) and marketed it through its Japan Victor Company (JVC). This JVC recorder had a marked advantage in that it could record for up to two hours. Because people who bought the recorders were using them primarily for recording feature films off-the-air, the two-hour format enabled them to record an entire feature on one cassette.[4]

With the advent of competition, the technological war was on and Sony altered its Beta format so that it, too, could record for longer. Then the succession of features was developed—pause buttons, timers, still frame, slow motion, high speed scanning, and taping methods that allowed for longer and longer recordings. Many other companies also began manufacuring cassettes with more of them adopting the VHS method than the Beta method. Zenith, Sanyo, Sears, and Pioneer were among those opting for the Beta format, while such giants as RCA, Hitachi, Sharp, Sylvania, and Magnavox manufactured using the VHS system.[5]

Once these videocassette machines became even semi-sophisticated, their sales soared beyond expectations. Sales passed the one million mark in 1979[6] and were up over three million a year by 1981.[7] Fotomat established a very successful business by renting videocassettes from its various film developing locations, and soon other videocassette rental and sales stores sprung up.

## Copyright Suit

Into this success story came a lawsuit. In 1976, a year after Sony introduced its Betamax, Universal Studios and Walt Disney Productions brought suit against Sony saying any device that could copy their program material was being used in violation of copyright and should not be manufactured. The case dragged through the courts and in 1979 a federal judge sided with Sony saying that copying TV program materials for use in the home comes under the "fair use" doctrine and is not an infringement of copyright.

However, two years later an appeals court sided with Universal and Disney saying that home recording was, indeed, a violation of copyright and that Sony, which, obviously, knew its recorders were being used for this purpose, was at fault. With several million recorders in homes throughout the country, this decision presented an interesting dilemma. No one advocated that federal marshals be sent into people's homes to confiscate the recorders. In fact, the courts did not prevent Sony from continuing to manufacture and sell more sets.[8]

Universal and Disney immediately sued the other manufacturers of both Beta and VHS recorders, Sony appealed the decision to the Supreme Court, which agreed to hear it in 1983,[9] and consumers bought videocassette machines and videocassettes at an even brisker pace worried that their source of a home movie library might evaporate.[10]

Fortunately for Sony and other VCR manufacturers, the Supreme Court, in 1984, reversed this decision and ruled that home taping does not violate copyright laws.[11] This decision was fortunate for consumers, too, for by 1984 over thirteen million recorders had been sold.[12]

### Rising Significance

By the mid-1980s, VCRs had become a fairly dominant device. Kodak and Polaroid introduced 8mm (¼″) VCRs that they hoped would cut into the market. Unlike the half-inch situation, a standard was set for 8mm making all machines compatible with each other.[13]

Movie studios began giving VCRs significant attention as a distribution method. They became the second entity to receive movies, preceded only by the motion picture theaters themselves. In other words, films were released to be dubbed onto cassettes before they were released to cable pay services, TV networks, or independent stations.

Discontent exists, however, regarding the method by which movie companies are paid for videocassette distribution. Because of an interpretation of the copyright law generally referred to as the "first sale doctrine," movie companies do not get a cut of the cassette rental business. They are paid only when a tape is sold, regardless whether it is sold to a video rental store or an individual consumer. Motion picture production companies and the people who might receive residuals, such as actors and writers, feel this method of payment is unfair. They feel they should be receiving a percentage from each tape rented.[14]

Rating companies have also begun reporting on VCR use. Nielsen conducted a study which found that the major use for VCRs was the recording of network shows, particularly daytime and evening soaps. They also found that 50% of people claimed they "zapped" commercials, either by fast forwarding through them or by using the pause button to avoid recording them. Saturday night was the most popular night for playing back programs recorded previously. VCRs also tended to be in homes of college-educated, high income people who watch more than the average amount of television.[15] Audience measurement technology exists which can tell whether or not viewers are fast-forwarding through commercials and whether a tape being viewed was made in the home or preproduced to be sold or rented in video stores. This technology is expensive, however, so has not been incorporated in the ratings methodology.[16]

Another problem to plague the videocassette industry is piracy. Estimates vary, but some think as much as $700 million a year is lost through tapes illegally dubbed and sold. Dubbing is a very simple process which can be undertaken by anyone with two VCRs. Techniques have been developed which render the audio inaudible or distort the picture when dubbing occurs, but these are not foolproof.[17]

Overall, the videocassette industry is thriving. Its future does not appear to be in danger, but its degree of success and its direction are uncertain.

# Issues

The major category of controversy surrounding videocassette recorders revolves around their effect on other media such as broadcasting, cable TV, and the film industry. Other issues involve the content of programs, standardization or lack thereof, and piracy.

## Effects on Broadcast TV

Broadcasters occasionally ponder whether videocassette recording is a friend or foe. Widespread use of VCRs would definitely dissolve the dominance of networks with regard to how people spend their time. No longer would people schedule their lives so they could watch their favorite programs. They could easily rearrange the schedules which network executives sweat and bleed over. But while videotaping might mitigate the power of the networks, it will also allow people who are not available at specific times to watch those network shows. In that way the shows might receive a larger audience than they would if people could not tape them.

Television programs are not the essential element, however. The programs exist primarily to deliver an audience for the commercials. The size of this audience determines the price that broadcasters can charge for the commercials. Advertisers are generally not enamored with videocassette recorders. First of all, the commercials can be ignored by fast forwarding through them when playing back the tape. Secondly, many ads are time related, and if people watch the programs a week or two later, the advertising is useless. For example, a commercial for a concert seen after the concert has occurred is of no value. Worse yet, many people record programs and then never watch them so never see the commercials.

On the other hand, if people watch the taped programs several times, or if they show the programs to friends, then the commercial may receive greater exposure than was intended. In fact, just the act of people watching a taped program they would not have watched otherwise means the commercial gets more exposure than it would otherwise. Generally, though, commercials receive the short end of the stick from videotape recording.

VCRs affect broadcast programming in other ways. If someone buys or rents a cassette movie, the time spent watching that no doubt subtracts from the time spent watching broadcast TV, a true reduction in ratings. Programs taped on Monday and watched on Tuesday prevent audience members from watching Tuesday's line-up of programs. When people are watching television as opposed to watching specific programs, they are much less discriminate about what they watch. If they feel like watching TV on Tuesday night, they will watch what is usually referred to as the least objectionable fare. In other words, from all the programs available, they will watch the one that they like best, even though it may not be particularly appealing. Many "least objectionable fare" viewers can increase a program's rating. But if viewers have as an option a program on a videocassette recorder, that will become their choice for the evening.

In an era when network audiences are being fragmented by cable TV and other programming sources that do not have time options, cassette recording further fragments in a manner that allows maximum time flexibility for the viewer. This could be very healthy for society, forcing an upgrading in programming. However, it could also reduce the economic base for the programming currently being produced, therefore, further lowering the quality.

## Relationship to Ratings
The size of the audience upon which the cost of the commercial is based is determined by ratings. In many ways VCRs play havoc with the time-honored ratings system. Generally, the program is counted as being viewed when it is taped, especially if the rating service uses an audimeter or other mechanical measuring device, which simply indicates what channel is tuned at what time in particular homes. In actuality, the program may not be viewed at that time at all. This raises the question of what type of audience composition should be credited to a program when no one is at home. If the rating service is using a diary method for which people write down the programs they watch each day, should the program be indicated in the diary at the time it is taped or the time it is watched.? How does a rating service calculate viewership at 9:00 P.M. Tuesday when some of the audience actually watches that program on Thursday? If a person watches the tape several times, should the rating be increased? Worse yet, if the person who tapes the program loans the tape to a friend, the rating service probably never records the viewership of the friend. That means the program will have lower numbers than it actually deserves, and loss of viewership numbers is loss of dollars. If a person tapes a program that is duly noted by an audimeter, but never watches the program, should numbers be subtracted from the rating calculations? If someone watches a taped program but does not watch the commercials, should the viewing be counted at all?

This last issue which involves the "zapping" of commercials is controversial in and of itself. Although audience measurement companies have found that 50 percent of people claim they delete commercials, these numbers are highly suspect in the broadcast industry. Generally, broadcasters and advertisers feel that people say they zap commercials because this is a socially acceptable response. No one really wants to admit that they watch commercials when they don't have to.

Although rating companies have begun to study VCR viewing, they have not refined their techniques. The whole area of VCR viewing is ripe for research of both an academic and audience measurement nature.

## Effects on Cable TV
VCR viewing affects cable TV viewing in the same way it affects conventional network viewing. If people are watching cassettes, they obviously are not watching cable at the same time.

VCR viewing has the potential of being more damaging to the pay cable services than to network TV in that much of the programming of pay cable and cassettes is the same—movies. Because cassettes receive movies before the pay services, they can skim off an audience that otherwise might wait to see a movie on pay TV. VCR's have been blamed for the fact that many people who used to subscribe to several pay services now only subscribe to one.

Some feel that VCR penetration will exceed cable TV penetration within several years, mainly because of the difficulties cable is having wiring the big cities. If people become accustomed to watching movies on VCRs before they have the opportunity to watch on cable, they may not bother to subscribe to any pay movie services.

## Effects on the Film Business

One area of the film business that has been hard hit by the emergence of videocassette technology is the area of super 8 equipment. Many people who used to shoot home movies with super 8 are now using cameras attached to their videocassette recorders for the same purpose. Sales have plummeted on home movie equipment while at the same time sales of consumer quality video cameras have increased.

The professional movie industry has a mixed relationship with videocassettes. On one hand, cassettes provide extra income to the production houses and distributors. On the average, the Hollywood entities receive about $1.50 per viewer when films are shown in a theater. When a cassette or a movie is sold for $50, Hollywould receives almost $6.[18] For this reason movie companies should be ecstatic about VCRs and attempt to turn business from theaters to videocassettes. However, most of the cassette business is rental for which the film companies receive nothing.

Hollywood lobbyists are trying to convince Congress to amend the copyright law to invalidate the "first sale doctrine". If this happens, production companies might release movies to cassettes at the same time they release them to theaters to capitalize on the cassette rental market.

Lacking this, film companies might try to arrange for cassette purchase prices to be lowered and rental prices raised. In this way, more people might buy the cassettes instead of renting them. This would increase the income potential for film businesses.

Movie theater owners would also like to see the price of rentals raised to at least the price of two movie theater tickets. Now, with rentals available for as low as one dollar, people can watch a movie at home much more cheaply than they can watch it at a local theater.

Because the cassette business is so dependent on movies, the film business will profit in the long run from its association with VCRs. But in the short term, the two will most likely continue to jockey for position.

## Program Content

Movies on cassettes are certainly successful, but other forms of programming are still floundering. The Jane Fonda exercise programs have been a big seller, but other types of self-help and educational materials have not been well developed. This is due mainly to the fact that not enough people buy or rent such materials to make them economically viable. As VCR penetration increases, such programs may become more widespread.

As with other new technologies, the abundance of sexually explicit programming available presents controversy. Cassettes have not been as violently attacked on this front as other media because purchasing or renting cassettes which contain such material is definitely an individual personal decision. People with cassettes can conceivably control their showing by locking them in a closet. They play the material fully aware of its content. They are not taken by surprise as sometimes happens with pay TV services.

However, the fact that this material does exist and is quite popular is opposed by many factions of society. Of particular concern is that it can sometimes be easily obtained by youngsters. Movie ratings are not taken into account in cassette rental, and sometimes children who could not see a movie in a theater because it is R-rated can rent a cassette of the same movie easily. Several groups, including the PTA, have urged that this practice be changed and that all cassettes be rated and sold or rented only to those who could see such a movie in a theater.

## Technical Standardization

The two different half-inch formats, VHS and Beta, cause problems. Masters of programming material must be made for both systems. Sometimes one system will acquire rights to material that is not given to the other system. Customers with the "wrong" system cannot have access to the program.

People who make home movies and want to show them on a relative's system often cannot do so if the two systems are incompatible. Tape exchange among individuals is also often thwarted by the different formats.

Because VHS outsells Beta three to one, there is some question as to whether or not the Beta format will survive, even though it is in some ways technically superior.

The 8mm format, although standardized, has not made significant inroads into the consumer market. The half-inch equipment seems fairly well established, especially the VHS format.

## Piracy

Videocassettes are extremely easy to copy. FBI agents have uncovered numerous professional style duplicating facilities totally devoted to the illegal transfer of cassette programming. The sex-oriented movies are particularly susceptible to this, although popular theatrical releases also have a large black market. In a few instances cassettes of major American films have appeared in foreign countries before they have even been released in the United States. One enterprising foreign pirate captured a film by sneaking a camera and videocassette recorder into a movie theater and taping the film off the screen. The technical quality was horrible, but he still managed to sell quite a few copies before he was caught.

The movie industry is very concerned about piracy and pushes for strong legal action and enforcement. There is no doubt about the illegality of copying films or tapes and selling them in a large scale way on the black market, but catching the culprits is often difficult.

The copyright law was established because creative people need just compensation. Creative works can be sold over and over again but still belong to their creator. Movies are sold to broadcasters, cable TV, subscription TV, and videocassettes, and yet through all these sales, the copyright belongs to an individual or a film production company. A company that manufactures videocassette machines or duplicates videocassettes does not have the right to permit someone else to make a copy of any material that it distributes. And most certainly a company does not have the right to duplicate and sell material without permission.

Taping that occurs by individuals in their homes is a much grayer area. Anyone who tapes a program off-the-air and then charges admission to others to see it is definitely violating the copyright law. But individuals who tape for their own time shifting benefit are definitely not considered pirates. Someone who buys a tape of a movie and then makes a copy for a friend, however, is on shakier ground. Even more nebulous is someone who tapes a program off-the-air and then loans it to a friend. These situations have not been tried in the court system yet because big companies are reluctant to pursue the "little guys" taping in their homes. Companies have, however, placed encoded material on tapes which makes them difficult to duplicate. The amount of money ostensibly lost to piracy is large enough that concerted efforts will continue to be made to solve the problem.

Although videocassettes do involve some controversial issues, their popularity is not threatened by any of these. Sales continue to be brisk and the industry is thriving.

# videodiscs

## Description

In some ways video technology parallels audio technology. The open-reel vid-eotape recorders have a sister in open-reel audio tape recorders. Both video and audio employ cassette recorders. The videodisc players can be compared to pho-nographs in that the software is contained on thin round discs and the machines can only play back, not record.

### Formats

At present there is only one format of videodisc machine—the laser disc. This machine can be connected to a conventional TV set in a manner similar to vid-eocassette recorders.

The discs themselves are twelve inches in diameter and made of a shiny silvery reflective material coated with plastic. They are placed in a disc player which contains a low-power laser where they spin at 1800 revolutions per minute. This laser reads the information without actually touching the disc so the discs have no grooves and never wear out. Discs contain thirty minutes of material on each side for a total of one-hour playing time. The material, however, is stored in in-dividual frames, 54,000 to a side or a total of 108,000 frames per disc.

The laser can access any particular frame quickly and accurately. A particular frame can be held in still frame almost infinitely, or various frames can be called up by number or name and appear instantly. This feature makes the laser disc particularly suitable for random access. For example, if a dictionary were stored on a disc, the definition for any particular word could be found quickly.

Some of the more sophisticated laser disc systems can interface with com-puters, thus expanding their interactive capabilities. For example, a furniture company might make a videodisc of its products, showing them from various angles and in various configurations. A customer could then interface this disc with a computer and order one or more pieces of furniture.

Laser discs can hold moving pictures, still pictures, or words all with equal ease. High speed scanning, such as that employed by the videocassette machines, is also possible on videodiscs as is slow motion or fast motion. The entire content of a disc can be viewed either forwards or backwards very rapidly or very slowly.

**Figure 12.1.** Pioneer's Laser Disc
video disk player.
Photo courtesy of U.S. Pioneer Electronics.

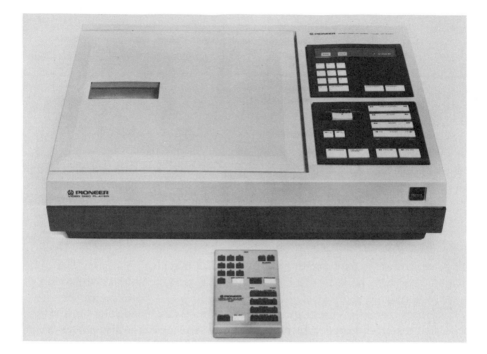

Laser discs have two sound tracks allowing the user to play very high fidelity stereo or to choose only one of the audio tracks. For example, material could be programmed in two languages, and the user could select the language preferred.

Some top of the line videodisc players can also play back compact audio discs (usually called CDs). These are 5¼" discs which hold several hours of digitally recorded sound—usually music. Video and audio discs can not be played back simultaneously, but the same machine can be used to play both.

Laser disc machines only play back; they do not record. However, engineers are working to develop a disc machine that will both record and play back.

Until 1984, a second form of videodisc, the capacitance electronic disc or CED, existed. It was much less sophisticated than the laser disc and more nearly resembled the audio phonograph record configuration. The machine contained a diamond stylus which rode over the grooves of the disc. The discs were enclosed in a plastic sleeve, and both disc and sleeve were inserted into the machine. The sleeve was then removed and the disc was ready for operation. When the user wished to remove the disc from the machine, he or she inserted the sleeve back in and removed the disc in the sleeve. In this way the disc was not hand touched, a factor which gave it a long life of several hundred plays. However, unlike the

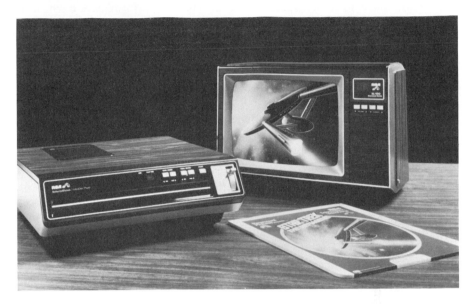

laser disc, it would eventually wear out because it was touched by the stylus. The capacitance discs spun at 450 revolutions per minute and contained two hours of programming, one hour on each side. Like the laser disc, it only had the capability of playing back, not recording.

The most significant difference between the CED and the laser disc was that the former did not have a multitude of features such as random access, still frame, or interface capability with computers. It could scan at high speed to find particular material, but it could not be directed to a particular frame. Because it did not still frame, there was little value in recording words or still pictures on capacitance discs. The main use for this videodisc system was the showing of program material, particularly movies. However, cassette machines did this just as well, and they had the added advantage of being able to record, so the capacitance disc never found a large market.

Videodiscs can transer to the TV set high quality pictures and sound, unrivaled by any of the video tape configurations. Many of their features exceed those of video cassette recorders, but, at present, they do not have the ability to record.[1]

### Programming

By now, it should come as no surprise that the primary programming of videodiscs, like most of the other new technologies, is movies. Disc companies line up distribution rights to films the same as everyone else, transfer the filmed material to a master disc and then press discs in a manner similar to the way phonograph records are pressed, through a molding process.

Because of the extremely fine audio capability of videodiscs, they have also been programmed to some degree with concerts and performances of musical groups. Games are also available on videodiscs and have been used in arcades as well as homes.

Although these entertainment forms were the primary programming of capacitance discs and have been the primary thrust of laser discs, the potential for informational programming on the laser discs is almost limitless. The interactive capability makes discs an excellent tool for educational material. Students can be presented with material and questioned about it. Depending on their degree of understanding, they can be advanced quickly through the material or placed on a slower track that explains the information in more detail. All the materials needed for the different levels of learning can be placed on the disc, but each student is led only to those which apply to his or her learning needs. In this way each student proceeds at his or her own pace.

Laser discs also have great potential for storage. The entire *Encyclopedia Britannica* can be stored on a few thin discs. This storage capability plus the ability to call up any particular frame enables users to find particular information quickly and thoroughly. Hospitals can store entire medical records, including x-rays on discs, and businesses can maintain files, financial data, and inventories. The equivalent of slide presentations can be stored conveniently because one disc is equal to 900 carrousel trays. This information retrieval is very rapid on laser videodiscs and can save a great deal of time for companies and individuals.

Because of the ability to combine movement with still frame, travel agents have developed video pieces which show resorts or hotels. A particular room can be held for detailed study. Similarly, material can be developed for a trip through the universe, a study of parts of the body, a tour through homes in the real estate market, or a view of the features on new model automobiles. With the branching capabilities of videodiscs, a person trying to select a hotel at a resort can choose to tour the weight-lifting facilities of several hotels, or a person studying the universe can zero in on particular aspects of a planet.

The military is using videodiscs to train people to use certain weapons. By programming a field of moving tanks and the characteristics of a weapon used to destroy those tanks, potential gunners can be taught firing techniques without actually using up expensive materials. These training techniques in some ways resemble video games, but the input is more sophisticated. In a similar manner, companies can use discs to train people in the skills they need for particular assembly line or skilled technician jobs.[2]

One department store is using videodiscs as a sales tool. It displays products on a disc and then, through computer interface, gives up-to-date price information.[3]

Two films, "Citizen Kane" and "King Kong," have been placed on discs with film sound on one audio track and commentary about the film on the other. The films can be viewed one frame at a time to aid analysis by film students.[4]

None of this informational programming is widescale yet, but studies are presently being conducted at various universities to determine the possible future uses of the laser videodisc technology. These studies and perseverance on the part of disc marketers and programmers could lead to exciting and innovative uses of the discs.

### Regulation

Like videocassettes, videodiscs are an unregulated industry with market conditions being the primary determinant of their future.

Unlike videocassettes, the videodisc industry is not likely to face legal battles over copyright. For one thing, the philosophy of the Supreme Court decision on the legality of video taping would probably extend to discs. In addition, videodiscs, at present, are very difficult to duplicate. Because they do not contain any provisions for recording, a disc played on one machine can not be copied onto another machine. The duplicating process is a complicated manufacturing process, involving much more than the simple output to input duplicating which can occur with videocassettes recorders.

Black market pirates are much less likely to set up the molding equipment needed to duplicate discs than they are to arrange a bank of videocassette recorders connected by simple wires. There is virtually no way people in their homes can make illegal or semi-legal copies of anything.

Overall, the videodisc technology would appear to have great potential. However, that potential has not yet materialized.


# History

The word that most aptly describes the history of videodiscs is "disappointing." One impediment after another has surfaced, making the viability of the entire video disc industry questionable.

### Early Development

Development of the disc began in earnest in the late 1960s and was soon followed by many pronouncements about the potential efficiency, low-cost, and flexibility of the disc. Announcement after announcement followed that the video disc would be on the market "next year," but many next years came and went unfulfilled.

The laser disc was under development primarily by a company called Disco Vision, which was formed by a union of MCA (the parent company of Universal Pictures) and N. V. Philips, a European-based manufacturing company. The original plan was that Philips would manufacture the disc players and discs, and MCA would provide the programming, primarily through its large library of films. Philips established a subsidiary called North American Philips. The subsidiary, in turn, purchased Magnavox, which was to be the distribution arm for the disc players. This alliance was not what would be referred to as a happy marriage, primarily because Philips was not able to work the bugs out of the disc technology.

Meanwhile, RCA was involved in the development of the capacitance disc. It developed the disc player and discs and sought programming from a variety of suppliers, primarily film companies. It also lined up a number of companies planning to distribute the hardware and software, including Sears and JVC. It, too, encountered innumerable technical difficulties that necessitated changing its projected marketing plans many times over.

## Initial Marketing

The first disc machine to actually hit the market was the MCA-Philips Disco Vision machine, which went on sale in Atlanta in December of 1978. The machines sold for $700 and discs varied from six dollars for a short to sixteen dollars for a full feature.[5] The machines sold out quickly, but then the problems began.

Disco Vision published a catalogue of 202 titles of program materials for its machine. However, only a small number of those were actually available; the rest were "soon to come." Enormous problems surfaced with the manufacture of the discs, so many of the planned features were never even mastered, and those that were mastered could not be duplicated properly. Although the manufacturing problems were numerous, the primary annoyance of the discs was that they stuck and could not be fast forwarded past the trouble point. People who watched movies could rarely finish them, a frustrating experience at best. There was nothing they could do but take out the disc and return it to the manufacturer, hoping for a better one in return. Disco Vision never caught up on the demand for discs even though only 25,000 machines were sold.

In addition to disc problems, Philips was not able to manufacture the disc machines in the quantity that had been anticipated, so although there appeared to be a demand for the machines, they were not attainable.

Philips and MCA had planned to begin marketing the machines on a city by city basis, but they cancelled those plans and went back to the drawing board.[6]

A year prior to the introduction of the laser disc, MCA had made a separate joint venture with Pioneer Electronics, a Japanese video disc manufacturer, which had produced players for the industrial market. MCA now began to tighten its ties with Pioneer and loosen them with Philips.

In 1979 MCA formed another partnership, this time with IBM, in order to gain additional needed capital and in order to use IBMs technical expertise to solve the disc manufacturing problems. The two companies set up a disc pressing plant near Los Angles.[7]

## A Second Try

In March of 1981 RCA began marketing its capacitance videodisc player called Selectavision. Instead of going on a market by market basis, it orchestrated a $20 million promotion campaign and went nationwide to 5000 dealers all at once, predicting it would sell 200,000 players in the first year. The companies tied with

RCA in this venture included Zenith, Sears, Sanyo, Hitachi, Toshiba, Montgomery Ward, and Radio Shack. The recorders sold for $500 and the discs for an average of twenty dollars each. RCA published a catalogue of one hundred programs, all of which were actually available.

The player price of $500 was less than the $700 that had been charged for the laser disc, but the technology and features of the Selectavision were not as extensive as the DiscoVision. The price was also about $1000 less than the videocassette recorders.

Twenty-six thousand players and 200,000 discs were sold in the first five weeks, and RCA was buoyed by its success.[8] The disc sales were far above what had been anticipated, and by stepping up production at the Indiana pressing plant, the supply was fairly well able to keep up with demand. RCA did not experience the disc pressing problems that had plagued its competitor.

**Selectavision's Demise**

However, as the months wore on, disc player sales lagged, and by October of 1981 only 50,000 players had been sold. RCA fell well behind its 200,000 projection. Executives at RCA stated that the disc players were not really having marketing problems, only that the prediction had been too high. They pointed out that in the first year of color TV set production, only 5000 sets were sold. They also glowed over the disc sales, which reached 2,100,000 in 1981, indicating that the average disc player owner bought thirty discs per year.[9]

RCA initiated an aggressive marketing strategy and lowered the price of its players to $200. Sales never picked up and by 1984 only 500,000 disc machines had been sold. In April of 1984, with only 500,000 disc machines sold, RCA reluctantly stopped manufacturing and marketing its disc recorders, blaming its demise primarily on the videocassette recorders which could record as well as play back. The company had invested $200 million in the development of the capacitance disc, very little of which it recouped.[10]

**Laser Disc Developments**

Meanwhile, problems continued to dominate the MCA-Philips-Pioneer-IBM disc entry. Pioneer was able to manufacture the discs in Japan, but the Los Angeles based facility never conquered its quality control problems. The primary stumbling block was dust. The plant was not able to attain the sufficient cleanliness needed to produce flawless discs.

In 1982 Pioneer bought the entire operation, taking both Philips and IBM out of the videodisc business. MCA remained involved as a program supplier only. Pioneer renamed DiscoVision, called it LaserDisc and set about solving the manufacturing problems.[11]

By the mid 1980s, most of the manufacturing problems of the LaserDisc had been solved, but the machines were still not selling well. Pioneer developed a machine that could play back audio discs as well as videodiscs and also began the development of a disc machine which would record as well as play back.

Because of the interactive features of the laser technology, the company was also aggressively selling to the industrial and education markets in hopes that the machine would be used for teaching purposes and for storage of information.[12]

## Issues

The primary issue facing videodiscs is viable survival. This involves such points as technical options, cost, program innovation, and historical failure.

### Technical Options

Now that only one format of videodisc is still being marketed, videodiscs do not suffer the incompatibility problem faced by videocassette recorders. If a disc format is to succeed, it will be the laser disc. Technical bugs seem to be worked out of this machine, and it does seem to operate reliably.

The main technical flaw seems to be that the disc machine can not record. Although this has advantages as far as piracy is concerned, it is a flaw that affects consumer acceptance. The main use of videocassettes is to record programming off-air for time-shift purposes. A machine incapable of doing this does not seem capable of being embraced by the public. Cassette recorders are now outselling disc players eight to one. By the time disc machines which can record are developed, VCRs may have saturated the market. Convincing people to own both a VCR and a disc machine may not be feasible.

In other areas of technical abilities, the disc outshines the cassette. Because the laser can move freely to any point on the disc, retrieval of information is rapid. A cassette tape must physically fast forward or rewind to a particular point—a slow process when compared with the instant access of a laser beam. Cassettes can not still frame indefinitely and, in general, are not valuable for anything that involves displaying one frame of information at a time. At present, the audio available with discs is superior to that of cassettes.

Yet cassettes are chosen over discs, mainly because of their capability to record.

### Cost

In the cost area, too, discs should theoretically have overruled cassettes. During most of their early history disc machines sold for prices which were several hundred dollars under VCRs. The gap has narrowed as VCR prices have come down to the $300 range, but on the average disc players are still cheaper than videocassette recorders.

The discs, themselves, were planned for a sale price of twenty dollars, under the assumption that videocassette tapes would rent for ten dollars and sell for fifty to one-hundred dollars. The disc prices have not changed, but some cassettes now sell for as low as twenty dollars. Worse yet for the disc business, tapes do not rent for ten dollars but rather for somewhere in the neighborhood of one to five dollars. This makes for a great price differential between renting a cassette and buying a disc. People who want to view a film once or twice would naturally opt for the cassette rental price. Rental of video discs has been tried on a limited scale, but given the handling involved, the discs can not rent for less than the cassettes.

The great cost advantage which advocates of the videodisc business espoused in the early days has largely evaporated due to success and tactics on the part of the videocassette industries.

Also, with the VCR, the consumer obtains more flexibility for the money because VCRs can be used to record "home movies" or off-air programming and also playback prerecorded material. The videodisc can only playback prerecorded material.

## Program Innovation
The real future for the disc seems to hinge on its ability to attract markets other than the consumer market. The disc has already lost out on the playback of feature films and to some extent on self-help programming such as exercise and how-to tapes.

Its strengths seem to lie in random access and interfacing with computers. Although there may be some applications for these traits in the consumer market, the main applications would come in business and education.

These markets, while willing to experiment with the disc, have not given it the wholehearted support that they have given to other technological innovations such as the computer. Before discs can be accepted, the videodisc industry must prove that it is more cost efficient and learning efficient to train people with discs than with live teachers, manuals, films, or other devices. Similarly, discs must be proven to be cost effective and convenient to industries such as real estate and travel which now utilize brochures, photographs, and, to some extent, videotape.

Perhaps new programming ideas, as yet unthought of, will prove the key to the disc's future. Because the disc has not been successful in entering its intended abode—the home, it must restructure its marketing techniques to fit into other applications. In a limited way, the disc has become a solution looking for a problem.

## Historical Failure
One reason the disc has not achieved great success is that the first discs malfunctioned. When the first units on the market did not operate properly, they left a bad taste in the consumer's mouth for the entire videodisc industry. In fact, RCA attributed some of its slow sales to the after effect of the intial laser discs.

Now that RCA has dropped its entire videodisc line, it once again gives credence to the failure aspect of the disc business and makes it harder for the laser disc to survive. A large, promotable success for the videodisc would be excellent medicine, but such does not appear on the horizon.

The videodisc is a high quality machine with many worthwhile features. If the industry can exploit its positive points, this technological device may yet become a success story.

# teletext

# 13

## Description

Teletext and its cousin, videotext, are both systems for displaying words, numbers, and graphics on the TV screen. They are new developments in the experimental stage which have the potential for becoming important information services.

### Over-the-Air Transmission

Teletext is broadcast over the airwaves in conjunction with regular TV stations by utilizing part of the regular TV signal. The creation of a video picture involves a process called scanning. Electron beams scan back and forth across the TV screen activating phosphorescent dots, which blink and glow in rapid succession. The TV picture, therefore, is not a picture at all but rather a succession of dots created by electron beams. The phosphor and the eye retain the image of the dots so that the effect of a total moving picture is created.

The electron beams signal the different colors of phosphor (red, blue, and green) to glow in varying degrees of brightness and at varying times dependent upon the picture that is to be created. For example, if the picture is primarily blue sky, an electron beam will activate the blue phosphors to a great degree while the reds and greens will hardly be activated at all.

An electron beam starts its scanning process in the upper left hand corner of the screen and proceeds to the right-hand side. Then it turns off momentarily, goes down slightly and retraces back to the left-hand side. Once again it scans from left to right, turns off, goes down further on the screen, and scans again. This process proceeds until the beam gets to the bottom of the screen. At this point it turns off and returns to the top of the screen to begin the scanning process all over again. This period during which the beam travels from bottom to top is called the vertical blanking interval.

It can actually be seen on a regular TV set by turning the vertical adjustment knob so that the top of one picture and bottom of another appear on the set. The black bar between the two pictures is the vertical blanking interval.

Part of this black bar contains information to keep the picture stable, but until recently, the rest of the interval was, in essence, empty. Engineers have now designed ways of using this empty or unused portion of the broadcast scanning

**Figure 13.1.** The TV scanning process. Actually 525 scan lines are placed on the front of the tube.

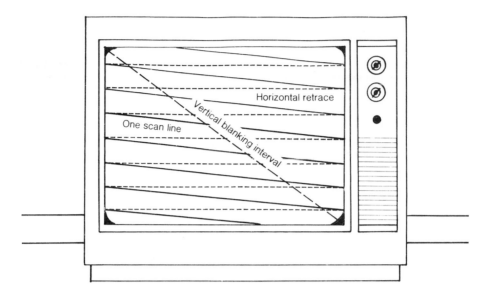

signal to carry information that will translate into words, numbers, and simple drawings on the TV screen. Teletext information, therefore, piggybacks on the regular TV station signal by using lines that are not used for video pictures or electronic processing.

## Reception

Although the vertical blanking interval can be seen on a conventional TV set, the words and numbers placed in that interval cannot be deciphered without a special decoder box. This decoder translates the electronic signal information to teletext and displays this teletext on the screen. Both the regular video programming and the teletext can appear on the screen at the same time, a process used for captioning television programs for the deaf.

Teletext information can also be seen instead of the conventional programming of a station. This is usually accomplished by a simple A-B switch attached to the decoder. In the A position the TV set displays the regular programming being broadcast by the station. In the B position the TV set ignores the whole video scanning process and instead displays the information coded into the vertical blanking interval.

One way teletext information can be displayed is in a simple sequential manner, one page after another. The viewer could, for example, place the A-B switch in the B position and watch what the station had pre-programmed, perhaps ten pages of news headlines followed by the weather and a stock market report. When the stock market report finished, the ten pages of news headlines would cycle again. The viewer would have no control over what page appeared when.

More sophisticated teletext systems allow for viewer choice through menus. The first thing that appears on the screen when a viewer turns on the teletext is a menu question, usually in multiple choice format, asking the viewer what information he or she would like to view.

The viewer uses a hand held keypad attached to the decoder to answer the question. For example, the teletext system might print on the screen, "Which information do you want: (1) News, (2) Stock Market, (3) Entertainment Available, (4) Transportation Schedules, or (5) Children's Games?" A person interested in the entertainment available would push number 3. The decoder box would receive an impulse from the keypad indicating the 3 and would then grab the information on entertainment as it cycled through the vertical blanking interval. Sometimes there is a wait of several seconds until keypad, decoder, and vertical blanking interval find the right information.

The first page on entertainment might be another menu to choose from such as: (1) Plays, (2) Movies, (3) Concerts, or (4) Dance Performances. The viewer who pushed the number 1 to select plays would then be shown a list of the plays being performed in town, perhaps with their starting times and ticket prices.

In this way teletext is slightly interactive. Viewers can select material they wish to view but only to the extent of the material that is programmed into the vertical blanking interval.[1]

## Cost

Teletext in this country is only in experimental stages, so the cost factor is an unknown quantity. One way for people to receive teletext would be for them to subscribe to it in much the same way that people sign up for subscription TV. They would rent or buy a decoder box and then pay a monthly fee for the service. A survey conducted by Oak Communications, a large decoder manufacturer for STV boxes, found that 76 percent of people contacted said they would subscribe to teletext and placed its value somewhere between five and twenty-five dollars per month.[2]

Another way for people to receive teletext would be to have the decoder built into the TV set. One study by a Salt Lake City TV station indicated that if fifty dollars were added to the cost of a TV set, 67 percent of the people would pay the extra amount to receive teletext. If the cost was raised to an additional $100 per set, only 42 percent of the people would buy it.[3]

This second method would need to be combined with advertising in order to cover the costs of producing and distributing the data. Advertisers have already expressed an interest in this medium. Ads would be more akin to newspaper and magazine ads than to TV ads in both cost and lay-out because they would consist of words and graphics. They could be programmed on separate pages that viewers could call up if they wanted to read that particular ad, but such a method would not be very acceptable to advertisers who would fear that viewers would not be likely to request their ads. More probable would be ads interwoven with information as now occurs in magazines. For example, the top half of the screen could relay baseball scores while the bottom half carried an ad for a sporting goods store.

### Programming
Teletext lends itself primarily to informational announcements: news, sports, stock market information, train and airplane schedules, emergency phone numbers, cultural events, consumer information, school bulletins, want-ads, and commuter information. One rather unique programming element, the programming of informational games for children, was tried experimentally by KCET in Los Angeles.

All of the programming must be designed so that it fits appropriately on the TV screen. News stories and other items must be written so that they do not end in mid-sentence at the bottom of the TV screen. Teletext is capable of a variety of styles and sizes of letters, so the screen arrangement is a matter of artistic design as well as letter counting.

### Graphics
The graphics created by teletext are not actual photographs or paintings but are more of a sketch or drawing nature. All teletext elements appearing on the TV screen are digitally produced so they are either on or off. The letters and numbers look similar to those on a digital watch, and the graphics are created by use of lines, squares, triangles, and other figures, which can be used to create visual representations. On some teletext systems color can be vivid and attention-getting, and limited animation effects can be generated.

Graphics are often used just to create interest, but sometimes they are pertinent to the content. For example, graphics can be used to show weather patterns or to show functions of an advertiser's product.


# History

Teletext is a new technology, which was developed in foreign countries, primarily England, France, and Canada, before it became a desirable concept in America. These three countries, who led the way, have background and experience, which the U.S. can draw upon.

### The English System
As early as 1966 British Broadcasting Corporation engineers began experimenting with ways to send information in the vertical blanking intervals of the TV signal. By 1969 they had developed a system, which they used experimentally to provide subtitles for the deaf and foreign language translations.

**Figure 13.2.** An example of teletext.
Photo coutesy of KCET, Los Angeles.

Then in 1972 the inauguration of two teletext systems was announced. One was a system called Ceefax (see facts) to be placed in operation by the BBC, and the other was a system called Oracle (an acronym for Optical Reception of Announcements by Coded Line Electronics) to be part of the Independent Broadcasting Authority, Britain's commercial TV network.

Both these used the same technology, but the BBC service, being government controlled, programmed information of public interest while the IBA, being advertising supported, tried to find the type of information that would be attractive to advertisers.

The technology of these two systems involved digital impulses piggybacking in vertical blanking intervals, a decoder, and a keypad, which the viewers used to select the pages to view. Ceefax transmits one hundred pages of information at a time, and Oracle transmits two hundred. The technology allows viewers to watch ordinary television programs with teletext appearing on the screen simultaneously. For instance, news bulletins can be shown over BBC programs. When the viewer no longer wishes to see the bulletins, he or she can use the keypad to erase them and then continue watching ordinary TV.

Material used on Ceefax and Oracle is stored in a computer. The various pages can be called up to a centralized keyboard located in the TV facility where they can be updated, eliminated, or reorganized. In this way the material can be programmed and reprogrammed easily. Graphics can also be created by converting images sent through a black and white TV camera to a special Ceefax unit, but overall, the graphics of the Ceefax system are very limited.

Britain's teletext service became popular slowly. By 1979 32,000 households in Britain had adapted their TV sets to receive Teletext, and by 1980 that number had grown to 120,000. Most people rent their TV sets in Britain. The added cost of renting a set capable or receiving teletext is about six dollars a month.

In 1978 Britain began serious attempts to export its teletext technology to other countries including the U.S.[4]

## The French System

The French teletext system called Antiope is used by Telediffusion de France, the government operated television network. It is very similar to the British system but was put into use somewhat later. In 1980 the French instituted a national program, the overall name of which is Telematique, designed to bring the combined technologies of computers and communications to every French household by 1995. Antiope is part of this overall program.

Like the British system, The French transmit text and graphics in digital form during the vertical blanking interval. This information and/or regular TV programming can be seen on the home TV. The primary program material is news, weather, and stock market reports.

The French, however, developed a method for producing higher resolution graphics than the Ceefax/Oracle system. For this system the decoders themselves contain character generators, which help generate graphics based on transmitted information.[5]

## The Canadian System

The Canadian system of teletext, first introduced in 1978, is called Telidon and is sponsored by the Canadian government. Most of the applications of it come under the heading of videotext because they do not use the airwaves. However, some teletext similar in nature to that of Britian and France does exist in Canada.

The Telidon system has the best graphics of any of the systems, but these graphics and other features make the cost of decoders much higher than it is for the English or French systems.[6]

## The Beginnings of American Teletext

As the British and French systems began developing in a fairly successful manner, a number of American broadcast stations began experimenting with teletext under authorization from the FCC. The first station to experiment was KLS-TV in Salt Lake City which tried the Ceefax system in 1978. In 1981 CBS's Los Angeles station KNXT and public TV station KCET both began experiments using the French Antiope system.[7] WETA, the public station in Washington, D.C., experimented in the same year using Canada's Telidon system.[8]

Because these various teletext methods were not compatible with each other, a conference of teletext experts and users was held in 1981 to determine if a standard could be set for the United States so that all teletext transmission would be the same. The British, French, and Canadian systems spent a great deal of time and money trying to influence American broadcasters that each of their systems was best.

But, AT&T appeared at the conference with a new teletext system called Presentation Level Protocol (PLP). AT&T's system was compatible with the French and Canadian systems because it incorporated the elements of those two systems and then added to them to make a more sophisticated system that outshined any others in terms of colors, letter variations, and graphics. The system was not compatible with the British Ceefax primarily in the area of graphics.[9]

AT&T and other American interests were hopeful that the FCC would designate PLP as the American standard, but the FCC chose instead to allow the marketplace to select a standard. Another conference was held in 1982 at which a slightly modified AT&T system called Presentational Level Protocol Syntax (PLPS) came on even stronger and gave the appearance of becoming the informal standard for the U.S. It was given the name North American Broadcast Teletext Standard (NABTS).[10]

But the British were yet to be reckoned with. They changed the name of their teletext system to World Standard Teletext (WST) and began an aggressive marketing campaign to convince teletext purveyors in the U.S. to use WST. While admitting that their graphics were not as realistic looking as the graphics of other systems, they pointed out that their system was far less expensive than the NABTS system and that it was already in successful operation.[11]

## Teletext Experiments

The following years saw a number of experimental teletext systems inaugurated, some with the British WST system and some with the AT&T NABTS system. In general, the start up costs for a station using the British system was about $30,000 and the decoders, manufactured by Zenith, cost about $300. Start up costs for the NABTS system were in the neighborhood of $150,000 and decoders were about $1000—when and if they were available.[12]

In 1982, Ted Turner's superstation WTBS began broadcasting teletext as well as its regular station programming, utilizing the British system. By 1984 it had leased some 300 decoders to cable systems, but no figures were available as to how many of those actually found their way into subscribers' homes. WTBS's suggested price for its decoders was a rental fee of $9.95 per month or an outright purchase price of $399.[13]

WKRC-TV in Cincinnati began teletext in 1983, also with the British system and a year later had sold 100 decoders for $300 each.[14]

KTTV in Los Angeles started its teletext operation during the 1984 Olympics to give Olympics results and traffic information. It placed decoders in shopping malls and other well-trafficked locations. After the Olympics were over, the station, which was using the British system, decided to continue teletext by programming news, sports, weather, job opportunities, ticket availabilities, bulletin boards, and quizzes. It is trying to place decoders in homes at $9.00 a month and is also trying to market the service to cable TV operators who can scramble it on an empty channel and charge subscribers $9.00 a month to receive an unscrambled signal.[15]

Experiments have been undertaken with the NABTS system, too. One of the first was conducted by Time, Inc. which delivered its service by satellite to cable TV homes in San Diego, California, and Orlando, Florida. This sytem started in October of 1982 and lasted only a year. During that time Time spend $30 million on the system and garnered about 250 subscribers. The project was dropped due to lack of economic viability. In addition, the decoders manufactured for the project turned out to be unreliable.[16]

CBS and NBC also launched NABTS teletext services in April of 1983. These national services are going to over 320 stations which pass them through to local areas. However, very few individual households receive the service because they do not have decoders. CBS has three stations—Buffalo, Los Angeles, and Charlotte, North Carolina—programming local material to go with the national service, and several other stations plan to do the same soon. However, the cost and unreliability of the decoders has prevented any type of large-scale marketing of the service. The best available decoder as of 1984 was one manufactured by Panasonic for $650. However, it only works well with a top of the line Panasonic TV set, so the cost to a consumer would actually be over $1000.[17]

Teletext can not be considered a success in the U.S. Probably no more than 500 people are receiving it in their homes. The providers of teletext hope to support it at least in part through advertising so that the cost to the consumer can be lowered, but so far advertisers have not been standing in line to add their messages to those transmitted on the vertical blanking interval. The fact that Britian has a million subscribers gives some hope to the American teletext industry, but at present teletext in the U.S. appears to be a solution looking for a problem.

## Issues

Something as young and immature as teletext has not had much time to build up adversaries, and yet there are issues that surround its existence.

### Standardization

From some points of view, a definite need exists to establish one standard for teletext in the U.S. If both the British system and the AT&T system become widespread, the consumer will suffer. Because the two are incompatible, a TV set manufactured to receive teletext transmitted by a station using WST may not be able to receive teletext transmitted by another station using NABTS. Likewise, if a consumer moves from one city to another, his or her TV set teletext converter, which worked in one city, might not work in another.

The problem with establishing a standard is that technology moves so fast the state of the art, when the standard is accepted, might be improved upon shortly thereafter. For example, if the FCC designated a particular standard for graphics, and all stations settled on that standard, a new invention might make possible even better graphics resembling photographs. Stations would not be able to take advantage of this improvement because they were locked into a standard.

This did happen in American television when the FCC decided that the number of lines from top to bottom on a TV set should be 525 lines. Prior to its decision the FCC had been asked to approve a 441 line system. It held out for better resolution, finally approving the 525 line system. But several years later, a 625 line system was developed and adopted by European countries. This system is superior, but the U.S. never adopted it because all TV sets would become obsolete.

The FCC does not appear likely to step in and standardize teletext, so the marketplace may have to decide which system survives. To some, that is the best approach. The American economic system applauds competition. Two systems competing should rule out lethargy on the part of either, and should lead to the development of a superior system.

Sometimes two systems find a way to coexist. For example, the two videocassette configurations, Beta and VHS, have survived for many years.

If both teletext systems develop temporarily and then one of them fails, consumers will experience minor inconveniences, but teletext does not represent anything life-threatening so the long-term negative effects will be negligible. This situation would be akin to the people who are left with defunct RCA videodisc machines for which there is no new programming.

Standardization is a thorny problem for teletext, but the overall concept of teletext will need to be more widely accepted before either standard can truly succeed.

### Technical Problems
The NABTS system must solve its decoder problem if it is to exist as a teletext system. Although it is technically superior to the WST system, especially in terms of graphics, it will never succeed if the decoders are unreliable. Unreliable decoders will leave a bad taste in people's mouths for the entire teletext industry.

As the technology of the NABTS decoder improves, its cost must diminish. This is quite likely to happen because the cost of manufacturing decreases once a product is debugged and can be manufactured in quantity.

But for the present, solving the technical problems surrounding the NABTS system needs serious thought and hard work.

### Programming Limitations
One problem with teletext is that the total number of possible pages in any present system is two hundred. When some of these pages are used for menus to ask viewers what material they want to view, the number of pages becomes even fewer.

The limited amount of information that can be provided may not be worth the extra cost the consumer would need to pay for the decoder.

On the other hand, two hundred pages may actually be too many, making the array of information confusing to the consumer faced with a lengthy complicated menu.

The type of information to place on teletext could also prove controversial. The tests involving preferred content are inconclusive. If advertising becomes the base supporting teletext, then the large corporations, heavy into advertising, may actually dictate what information is placed on teletext. Advertisers will want to highlight their information by placing it conspicuously in material that is likely to be read often, perhaps more for entertainment than information reasons. If advertisers gain this control, then the public service route, which teletext seems headed for at present, could evaporate.

## Competition

Teletext is going to face strong competition from other media such as videotext, which have greater flexibility. Teletext is one-way only. The consumer cannot send information back to the TV station to ask further questions or request particular information. Systems with which it will compete have two-way capability.

Other systems can also provide more complete and thorough information than can be supplied by teletext and can provide this information much more rapidly than the teletext system. The consumer truly interested in information access may find that teletext does not fill the bill.

## Sociological Effects

Overall, teletext does not face as many sociological issues as many of the other new technologies. Neither its content nor its delivery method lend themselves to some of the crucial issues such as piracy, privacy, and pornography. Teletext might lead viewers to sit in front of their TV sets extra hours per day, but handling teletext involves as least minimal action and thought on the part of the consumer.

# videotext

# 14

## Description

Videotext is related to teletext in somewhat the same manner that pay cable is related to subscription TV. To the uninitiated viewer, the two may appear to be identical when seen on the screen, but the methods of transmission and many of the features are different.

### Comparison to Teletext

While teletext is delivered over the airwaves using the vertical blanking interval of regular broadcast signals, videotext is delivered over wires, most generally cable TV wires or telephone wires. This, in essence, is the primary difference between videotext and teletext; the former comes over wires, and the latter comes through the airwaves.

This difference causes another main difference between the two technologies; videotext is two-way, and teletext is one-way. Because people cannot have broadcast transmitters in their homes, they cannot send signals back to the TV station requesting particular information. When using teletext, they can merely select from the pages of information sent out by the broadcaster, those pages they wish to view. This selection process occurs when the viewer presses a particular number on a keypad, which signals the decoder to catch a particular page being transmitted in the vertical blanking interval.

### Two Way Wire Delivery

Because videotext is delivered by wire, information can travel both from a source to the home (downstream) and from the home to the source (upstream). The sending source is usually a large computer filled with information. This information is sent from the computer to a device called a modem, which encodes the information into a form that can be sent over phone lines. These phone lines then bring this encoded information to the home telephone, just as they bring audio portions of phone calls.

The phone in the home can be connected through another modem to a simple computer called a terminal. The modem decodes the information so it is understandable by the terminal and the terminal then generates the letters and numbers which appear on a TV screen.

**Figure 14.1.** Videotext process.

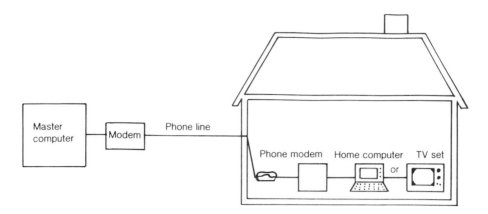

A more sophisticated home computer can be attached to the modem in order to generate letters and also store, process, or recall information.

By using the same computer-modem-TV set equipment, the consumer can send messages upstream to the large sending computer. For example, the user, who wants to buy a jacket, could type into the home computer a set of numbers that would access from the large computer something equivalent to the Sears catalogue. The TV set could display the numbers typed so that the consumer knows that they were typed correctly. The home modem encodes the numbers so they can be sent through phone wires to the large computer's modem, which decodes the numbers and sends them to the large computer. Using this same network (computer, modem, phone lines, modem, computer, and TV set), the user can obtain information about jackets and select a particular jacket to buy. When this jacket is selected, the user can type in specific information such as size and color desired, and in essence, order the jacket from the large computer. The large computer then sends this information (by using modems and phone lines) to a computer at the company that sells the jackets. When the company receives the sales order, it can mail or in some other way deliver the jacket to the consumer. This type of interaction could not take place with teletext because of the inability of the consumer to communicate to the source of information.

When two-way interactive computer oriented videotext operates through cable TV the process is similar except that instead of employing phone wires, which use audio sending techniques, the system uses the cable company's coaxial cable system, which operates by using radio frequencies. Some of the cable is designated for downstream transmission of both video and videotext information, and some is designated for upstream transmission of videotext information. The telephone is not involved in the cable TV process, but various devices still interface the home computer with a large computer. This large computer can be located at the cable TV's headend or the cable company can subscribe to various established computer information services and merely relay the information to the home consumer.

If this form of TV-computer interaction becomes more prevalent, people will view their TV set in a different light. It will no longer be a passive device bringing in information and entertainment but rather an active device over which the viewer exerts control and desire.

## Simple Videotext

Not all videotext uses are as complicated as the jacket buying example given above. Words and numbers can be delivered by phone wires or cable wires without full blown two way interaction. In fact, sometimes the term videotext is used to encompass all letter and number services utilizing wires, and the term viewdata is used to indicate those services that are highly interactive, such as purchasing items or conducting banking transactions. In other uses, the term videotext encompasses all text services (both through the air and over wire) while teletext refers to over-the-air and viewdata refers to interactive wire services. This designation leaves no word for the non-interactive cable TV text services.

Videotext, most commonly known as all text services coming over wire, can operate in a manner similar to teletext in that it can display a menu of information available to the consumer from a central computer. This information could include news, weather, want-ads, school bulletins, and all the other public service information associated with teletext. By using a keypad or a home computer, the consumer can request particular information displayed on the menu.

If the consumer uses the keypad to request choice 3, an impulse representing the 3 is sent through the coaxial cable to the computer at the cable TV headend. This computer then sends back the appropriate information to the TV set located in the subscriber's home.

The information does not ride on the vertical blanking interval; it appears in a manner similar to a regular TV picture. Also, the information available is not limited to the 200 pages that can be carried in the vertical blanking interval; the quantity of information is limited only by the size of the master computer and the ingenuity of the people gathering, programming, and revising the information.

On an even simpler level, cable TV companies can simply display videotext material over one or more channels. By using a character generator at the cable headend, someone can type in messages of interest, which cycle on the TV viewer's screen one after another all day long. Some cable systems have one channel devoted to community announcements. Twenty different announcements of community activities or events may be typed in and cycled all day long. Each morning someone updates the announcements eliminating events that have passed and adding ones planned for the future. In a similar manner, cable systems can designate channels to the weather, news headlines, traffic information, or upcoming programs on the cable system's access channels. Viewers wanting to find out about the weather simply turn to the weather channel and see the information they want without the use of keypads, decoders, modems, or home computers.

## Videotext Standards

Standardization of the textual signals (both videotext and teletext) revolves around such factors as intensity of color, form of graphics, and flexibility of lettering, and not around method of delivery. Therefore, the same standard can be developed for both videotext and teletext. At various times the British Ceefax, French Antiope, and Canadian Telidon systems have done their best to woo the videotext suppliers as well as the teletext purveyors.

The conferences to establish standards and discuss industry progress encompassed both teletext and videotext. In fact, the company that proposed PLP, the system that became the basis of the North American Broadcast Teletext Standard, was AT&T, which, because it deals with phone lines, is involved with videotext rather than teletext. At present, the videotext industry, like the teletext industry, is split between the British WST and the American NABTS standards.

## Costs

Data retrieved by use of phone lines is already in place. Several services such as The Source and Compuserve have large computer data banks of information available. Anyone who wants to invest in a home computer (such as an Apple II or a Radio Shack TRS-80), and a modem can access the information in these data banks. The pricing varies according to the time day, but individuals who use the system at night generally pay about five dollars an hour. At present the information is geared more to businesses than individuals, and those businesses accessing during normal business hours pay more. Any accessing of the system is billed in a manner similar to telephone billing.

The very simple cable TV videotext services such as cycling program schedules and community bulletin boards are often included in the very lowest tier of cable service. Anyone who pays the hook-up fee and the cheapest monthly fee receives these channels as part of the package.

The more sophisticated two-way services are sometimes charged on an as-used basis that is competitive with the phone service of about five dollars an hour. Sometimes these services are billed on a monthly basis for about the same cost as a pay cable service—ten to fifteen dollars a month. Most of the highly interactive services are in the experimental stage, however, so have not established a fee structure. They are offered free or at a very nominal rate to those households chosen to participate in the experiments.

Advertising is in the wings for videotext much as it is for teletext, but the line between advertising and selling can become blurred. A company selling jackets will want to present those jackets in the best light, but such advertising techniques can also lead directly to point of purchase sale not possible through conventional ads.

Advertisers may also be able to defray the cost of some of the informational retrieval, making its cost lower to consumers. For example, one type of information stored in data banks is recipes. If all recipe retrieval were partially underwritten by a large food company such as Del Monte, the Del Monte name or logo could appear in graphic form on each recipe.

Other information such as sports scores, weather, and stock market reports could have sponsor identification throughout or at the beginning and/or end of the material.

### Programming
Imaginations can go wild in creating uses for videotext, especially the highly interactive viewdata. Ideas thought of and, in some cases, tried, include banking, paying bills, ticket sales, public opinion polls, credit checks, games, at-home shopping, voting, homework help, meter readings, and adult education.

If a printer is attached to the TV set so that what appears on the screen can be printed out on paper, then the possibilities become even larger. "Newspapers" could be delivered every day through the TV set. All the data in libraries could be accessed, and those bits of information desired could be printed out. An entire lending library of books and magazines could be delivered by videotext. These could be read off the screen or printed out and taken to the beach. Mail could be delivered electronically by going from one home computer to another. Even "junk mail" could be sent from centralized computers to individual homes. Information people wanted to keep could be printed out or stored on a computer floppy disc. The rest could be discarded without even taking up any room in the waste basket. The TV set and printer could act as a glorified copying machine making multiple copies of information stored in computers.

In addition, videotext can show either on a cycling basis or by specific request news stories, weather reports, telephone yellow pages, stock market reports, community calendars, and all the other vast array of material of interest to the public.

## History

As with teletext, videotext was developed in foreign countries, primarily England, France, and Canada. The innovations of these countries led to further innovations in the U.S.

### The English System
The technology used for the British videotext system is the same as that used for the BBC teletext Ceefax and the IBA teletext Oracle. However, the name given to the British videotext is Prestel (short for press and tell). It is run by the British post office and was conceived in 1971 by a research engineer, Sam Fedida. He demonstrated the system to the managing director of the post office in 1974, and by 1977 the system was available to the public. It utilizes the phone lines so is a telephone-television hook-up.

The services offered by Prestel include travel turntables, weather, news, sports, entertainment and economic data, stock exchange information, food prices, children's games, farm news, want-ads, and business news. In all there are over 25,000 topics divided into subjects and divisions of subjects. Although Prestel was designed with the home consumer in mind, it has been used primarily by businesses.

This might be due to its cost, which is two-fold. First, it costs approximately twenty-five dollars more per month to rent a TV set equipped to receive Prestel than a TV set just equipped to receive video programming. Secondly, the cost for accessing Prestel is determined by the information provided. Some pages such as advertisements are free, but other pages cost from fifteen cents to one dollar. The cost of each page is shown on the upper corner of the TV screen, and the total cost of an entire session is similarly displayed for the consumer at the end of a session.

Much of the information provided by Prestel actually comes from outside providers such as corporations and non-profit entities. They determine the price they wish to charge for their information.

In order to use Prestel, a person places a local phone call to a Prestel number by pressing a special button on the keypad connected to his or her TV set. The call is placed without having to actually use the phone beause the modified TV set has its own phone jack connector. The call is routed to the nearest Prestel Information Retrieval Center, and the customer is charged for a local call.

The user enters a password, which indicates to the central computer that this person has authority to access information. Then, by pushing appropriate numbers on the keypad, the user retrieves the desired information.

Because Prestel is a two-way system, it has been used for sales. Companies create order coupon pages, which consumers can fill out using their keypads. The completed coupons are stored in the master Prestel computer and accessed by the companies that created them. The customer's charge card number is billed, and the order is processed.

Prestel, the oldest of the videotext systems, is growing slowly at the rate of about 700 subscribers a month.[1]

### The French System
The French videotext system utilizes the same technology as the French teletext system, Antiope, but goes by the name of Teletel. Teletel began its first experimental trial in March of 1981 in several French towns. It soliticited 2500 volunteer homes connected to a videotext center.

Several different types of tests were run. Some homes had facilities to shop at home, make travel and hotel reservations, view timetables, read the latest news, play computer games, and receive computer aided instruction in several topics. Another trial involved sixty farms that were supplied with the latest commodity prices.

Other trial participants were given an electronic phone book of both white and yellow pages. Using terminals they were given free, they could type in the names of a person, business, or service they desired, and the appropriate phone number or numbers appeared on the TV screen. The French plan to replace printed telephone directories with electronic ones. In that way, they can update phone number changes constantly, and the government will not need to undertake the expense of printing phone books.

Another French experiment involves the use of Smart Cards. These are slim credit cards with built in microprocessors, which connect to Teletel home terminals. By using these, participants can bank, transfer money, and pay bills.

All of these videotext experiments utilize phone wires. Future plans call for experiments that print out written documents.[2]

## The Canadian System

The Canadians have given the same name, Telidon, to both their videotext and teletext systems. The videotext is sponsored by the government but operated by Bell Canada utilizing phone lines. The Canadian system, like the French and English systems, consists of data in a centralized computer, phone lines, a TV set, a terminal, and a keyboard.

Canadian experiments with Telidon have been primarily in the field of education. Course work has been enhanced by information from Telidon, and the graphics have been used to build forms that students try to recognize.[3]

At present the Canadian system is very expensive with terminals costing about $1000. As a result, few individuals are utilizing the system. Some companies and the Canadian government iteslf are the main users. One of the activities the government has sponsored is the placing of 100 terminals in public locations where they can be used by all interested persons.

## Early Text Uses in the U.S.

Some of the forerunners of videotext existed in cable TV before the term, videotext, was coined. In the early 1970s character generators were developed, which were essentially typewriter keyboards, that could type material directly on to a video screen. Once these were operational, cable TV systems could type announcements, which could be sent over the cable on one of the empty channels. Some cable systems utilized this technology for program guides, weather reports, public service announcements, commercial messages, and the like. Several wire services adapted the material they were sending over teletype machines to the format needed for the TV screen and provided nationwide news story services, which cable companies could offer to subscribers.

In 1977 when Warner Amex began its Qube service on its cable TV system in Columbus, Ohio, it made use of text and two-way capability. The Columbus subscribers who participated, first on an experimental and then on a paying basis, were given keyboards with buttons that could be used in a multiple choice fashions. These keyboards were connected to the cable system so that the output from them could be sent upstream to a computer terminal at the cable company.

The first uses of Qube were of a public polling nature. Multiple choice questions were typed by the cable company and sent to the TV sets of Qube participants. The viewers read each question and pushed the button corresponding to the answer they chose. The impulses from the buttons traveled to the central computer where they were tabulated. The percentage of people giving each answer was reported by the computer and shown to all participants on their TV sets. In this way, Qube subscribers could know immediately how their answers related to those of their neighbors.

Some uses of this Qube system were just for fun. During a football game Qube would ask viewers what play the quarterback should call. Viewers could give answers, learn what others thought, and then see what the quarterback actually did.

Other uses of Qube involved opinion polling on local issues such as street pavings or park locations, voter preference polling, and product testing.[4]

Neither the character generated announcements nor Qube were called videotext and, indeed, neither of these initially employed the more sophisticated viewdata uses of videotext. The interest in videotext was aroused for several reasons, one of which was the experiments conducted in foreign countries in both the teletext and videotext areas.

Another element that set the scene for videotext was the rapid growth of cable with its interactive possibilities. As companies competed for franchises, they began promising the "mostest and bestest" in the state of the art of technology. Fancy shopping and banking services appealed to city councils, and as companies began fulfilling their franchise obligations, they expanded the capabilities of videotext.

Another catalyst was the growth of the computer industry, which made compiling and distributing large amounts of data on an as needed basis feasible. Computer data banks were established and became a logical source of information for videotext.

## Videotext Competition

Initial experiments with videotext both in this country and others were positive enough to make the concept look exciting. Suddenly, a large number of companies began considering the lucrative profits that might be available from this technology once it became developed. Competition rose as various entities tried to establish primacy in delivering textual information to consumers and businesses.

Newspapers felt they had reason to place videotext in their domain. After all, it amounts to electronic publishing, and they are the kingpins of the publishing business. If people were going to have news delivered to them by written word, the newspapers were the proper providers. Newspapers were also looking for a new frontier. Newspapers sales had shrunk enormously since the advent of TV newscasts, and many newspapers had consolidated or gone out of business. Rather than sitting back and allowing TV to further erode their readership, they were striking out to bring the action to their doorstep.

**Figure 14.2.** A sample videotext page.

Photo courtesy of Group W. Cable, Buena Park, CA.

It is not surprising, then, that two of the companies experimenting most heavily in cable delivered videotext are Knight-Ridder Newspapers, Inc. and Times-Mirror.

Knight-Ridder is operating a system called Viewtron in southeast Florida. It began in October of 1983 and by 1984 had approximately 2500 subscribers. It utilizes a terminal manufactured by AT&T which sells for $600. Subscribers pay an additional $12 a month for the information which includes banking, shopping, news, encyclopedia access, and information about the local area.[5]

The Times-Mirror system, called Gateway, began operation in 1984 in southern California and operates very much the same as the Knight-Ridder set-up. Neither of these systems have been overly successful in terms of subscribers or revenue, but they are still young and have shown some signs of growth.[6]

Cable companies feel they should control the videotext area because they have the necessary cable delivery mechanism and are, themselves, in the video business.

Group W began a service in 1984 at their Buena Park, California cable system. This operates very much like teletext in that subscribers, through the use of a keypad, can "grab" specific pages. The system has on it material that was selected from a number of different data services available through satellite and land lines. Group W also originates some of its own local material. One feature is a shopping service, but this is not cable interactive. Subscribers wishing to buy the items shown must use the telephone to place orders. The overall service, called Request, is available only to Buena Park cable subscribers for $7.95 a month.[7] In a similar manner Cox Cable and ATC have experimented with videotext services.

Another company which has asserted itself in the area is AT&T. As owner-operator of the largest phone system, it has felt it should have jurisdiction over text services which it, of course, feels are best delivered by telephone wires. Neither the cable companies nor the newspaper companies took kindly to the idea of AT&T's entry into the field mainly because it is too big and powerful. They feared that AT&T, because it already has long lines in place, could absorb the videotext field before others got to the starting gate. Whether videotext is to be delivered by phone or cable the long distance lines are needed to bring information from the main computer to the cable headend or to the individual phones. Because AT&T owns these long distance lines, newspaper publishers and cable companies feared it could favor itself in both price and convenience in delivering this information. As a result, newspapers and cable lobbied againt its entry and were at least partly successful through a rather complicated legal decision.

In 1956 a consent decree was issued that barred AT&T from entering non-regulated fields such as videotext. It could act as a conduit for informational services but it could not create or own them. In January of 1982, a long standing antitrust suit was settled that changed the rules governing AT&T. At this time AT&T and the justice department decided that AT&T should divest itself of all its local phone companies—a total of two thirds of its assets. AT&T could keep its long distance lines, its equipment manufacturing capability, and its research oriented Bell Labs. It appeared that AT&T was now free to enter into new telecommunications ventures such as videotext and, perhaps, even cable TV.[8]

But the details of the agreement were to be worked out by U.S. District Court Judge Harold Greene. The newspapers presented their case forcefully, and in August of 1982, Judge Greene modified the agreement to prevent AT&T from entering the electronic publishing business for at least seven years. AT&T, eager to settle its long standing problem with the government, agreed to this restriction. The exact interpretation of "electronic publishing" has not been determined, but the newspaper publishers and cable TV operators are very happy with the judge's decision.[9]

Videotext is at best a fledgling industry, weak on actual experience in the marketplace. Whether or not consumers and advertisers will feel its services are worth paying for has yet to be determined.

# Issues

Because of the insecure, uncertain nature of the future of videotext, many issues face this technology. A few of these, such as the need for technical standardization and the possible adverse effects involved with advertisers determining content, are similar to the issues confronting teletext. But because videotext with its two-way nature has the possibility of being more complicated and more encompassing, it has many issues unto itself.

## Economic Viability

Videotext is an expensive proposition. It involves the purchase of a home computer or the costs of modifying or adding to a TV set so that it can receive the text material. Added to this is the cost of the material itself. Although the industry has not decided as yet whether to charge by the hour, the page, or the month, a recent market research study concluded that by the end of the 1980s, the average household would be paying seventy-eight dollars per month for text services.[10]

Although the firm conducting the study felt consumers would be willing to pay this amount and that by the end of 1991 virtually all homes would have text capability, this has yet to be proven. Seventy-eight dollars a month is a substantial fee to pay for information and services that are available in other cheaper, albeit more time-consuming, ways. Just how much people will be willing to pay for the convenience of buying a jacket in their home as opposed to driving to the local store has still to be determined.

Of course, the possibility exists for advertisers to pick up some of the cost. Companies selling products and services could be charged for placing their information in the central computers. Other companies could be charged for having advertising material incorporated into the pages that the user retrieves from the videotext system.

This will only be attractive, of course, if the advertising pays off and in one way or another brings the company added business. Right now with videotext in the experimental stage, the cost per sale is enormous, but that will be reduced if videotext becomes widespread.

In order for videotext to become widespread, it must be priced to please the consumer. This chicken and egg problem of having the price low enough to attract enough subscribers to actually lower the price can be solved by some large company subsidizing videotext while it gets off the ground. Several companies have toyed with this idea but have backed off. If videotext is perceived as having such enormous long term possibilities that the short term is worth sacrificing, and if healthy competition exists among companies wishing to establish themselves in the field, then the development period for videotext may be short and consumers may get more than their money's worth at an early stage. If this does not happen, videotext may languish on the vine.

## Competition

Competition exists not only within the videotext area but between videotext and other new technologies. Teletext, because it does not require all the home interactive equipment, can be cheaper than videotext and may be enough for most people. Videotext is a more complicated system that involves use of a computer and knowledge of or reference to various codes needed to access material. Teletext merely involves a keypad, which can be operated in multiple choice fashion. People who simply want to know traffic conditions in the morning may feel more comfortable obtaining the information from teletext than videotext.

Videodiscs can contain large banks of information that can be accessed without contacting a central computer. Books, magazine articles, recipes, and other non-changing information might be more conveniently stored on video discs. A combination of teletext for changing information, video disc for stable information, and trips to the local shopping mall for shopping, banking, and other activities, might be more attractive to the consumer than videotext.

Within the videotext area itself, competition is keen among various companies eager to enter and control the field. Even though AT&T has been barred from direct participation in electronic publishing until 1990, it is still involved because it owns the phone lines over which much of the material may travel. The cable and newspaper companies vying for control could so fragment the market that none of them make a profit. This could delay the development of videotext and also open the door wider for AT&T.

The arguments over whether videotext is best delivered by phone lines or cable TV is unresolved. Almost everyone has a telephone, while only about 40 percent of homes have cable TV, so the phone system is more in place. However, the cable industry is geared more toward video and has been providing material of an informational as well as entertainment nature.

## Copyright, Fraud, and Obscenity

Because videotext can be considered electronic publishing, it will suffer some of the same legal problems as the publishing industry. If people call up a magazine article through videotext and then print that magazine article on the printer attached to the TV set, have they violated the copyright law? If they sell the copy to a friend who does not subscribe to a videotext service, have they further violated the law? During the rise of radio, broadcast TV, and photocopying, the U.S. lived under a copyright law written in 1919. This caused innumerable problems because the law did not cover applications of these new technologies. The law was rewritten in 1976 to take those inventions into account, but, obviously, it did not cover the even newer technologies such as videotext. The potential for copyright problems with a process that can occur in the privacy of a home is enormous and will need to be dealt with in both Congress and the courts.

If mail is delivered electronically through videotext, what is to prevent unscrupulous individuals or companies from offering get-rich-quick schemes, which are fraudulent. At present laws exist against sending fraudulent material through the mail, but even they are hard to enforce. The U.S. Post Office is an arm of the government so is fairly free to prosecute fraudulent mail. Videotext is planned as an unregulated industry operated by private corporations. They may not be interested, let alone able, to help prosecute such deception. The speed of videotext can also cause problems. Companies could promote fraudulent schemes, receive instant reponses from the most gullible, make sure the money has been properly transferred to them, and then disappear—all within an hour or two.

The possibility for the transmission of obscenity also raises problems. People may come home to find obscene messages in their electronic mail. As with fraudulent mail, how is this prosecuted? Some books and other materials considered obscene, at least by large segments of the population, may be part of the data in computer banks. Who is to decide what should and should not be there. Children who have access to a data bank to help with homework will be able to type a word such as "sphinx" on their home computer and receive a wealth of information about that subject. What is to prevent those same children from typing in "_____" and receiving other, more inappropriate information.

With copyright, fraud, and obscenity, the problem of liability arises. If, indeed, copyright violation can be proven, who is responsible; the person who created the material, the data bank that stored it, the phone company or cable TV company that transmitted it, the person who copies it, the person who bought it, or some combination of the above? If people are bilked fraudulently and cannot find the actual source, to whom can they turn? If anonymous, obscene mail appears, what will be the procedure for having it stopped? Should the company delivering the mail be in the position to screen (censor) all that it transmits?

These and many more as yet unthought of problems are likely to arise with videotext. They should keep the legal industry busy.

### Privacy
Of all the issues facing videotext and other two-way technologies, the one of privacy is the most publicized. The American style of life values the individual and that individual's privacy. Threatening privacy is bound to ignite public indignation.

All actions undertaken with videotext would be recorded in a central computer. This computer and those who have access to it could know everyone's purchases, income and outgo, leisure habits, reading patterns, vocational and avocational interests, and much more.

This becomes particularly scary in light of the knowledge that during the McCarthy era library check-out records were subpoenaed in order to expose subversives. If this type of witch hunt occurred again for any reason, videotext data would be much easier to come by. If a cable TV employee could combine knowledge of purchases with knowledge of home security, that employee might be able to orchestrate a robbery. Some totalitarian governments have concluded that the easiest method of keeping track of individual citizens is to keep track of individual purchases. Constant knowledge of what a person is buying and where he or she is located at the time of purchase makes knowing the whereabouts of everyone an easy chore.

Americans live in a society where privacy abuses are rare, and yet the possibility exists. Setting up a mechanism to control videotext data so it is not used to invade privacy is difficult.

Warner Amex did develop a privacy code for its Qube system. This code stated that individual viewing and response information would be recognized only where necessary to permit billing or render a subscriber service. For the most part responses would be lumped together and reported as percentages. This bulk data could be made available to third parties as long as the identity of individual subscribers was not ascertainable from the data provided. If any reports involving material about individual responses was to be published, subscribers had to be advised in advance and given adequate opportunity not to participate. Mailing lists could be made available to others but not without first providing subscribers the opportunity to have their names removed. In addition, any information gathered by Warner Amex pertaining to subscribers would be made available to them on company premises for examination and copying. The code also stated that Warner Amex would not release information about individual subscribers to government agencies unless a legal document such as a court order or subpoena is supplied.

This code, and others like it that may be drawn up by other companies involved with two-way communication, acted as a noble attempt at controlling privacy, but once the information is assembled in a computer, nothing is fail safe.

### Effects on the Poor
If videotext does turn out to cost approximately seventy-eight dollars per month, and if it becomes a primary source of information, then the poor will suffer. Those who do not have the money to subscribe to videotext will have to get their information through more cumbersome, time-consuming methods. This will widen the gap between the information rich and the information poor.

If videotext's two-way capability becomes used extensively for public opinion polling, the ideas of the poor may be very underrepresented. One criticism leveled at the poll taking of Warner Amex's Qube system was that the participants were primarily upper income and conservative. Reports issued, which gave results about who was most likely to be the next president or how people felt about certain economic policies, were taken seriously and had the potential to affect policy and opinion even though the reports stated that the results were from Qube subscribers. People with videotext could become more influential than those without, and people with videotext will be the more affluent.

### Social Ramifications
In fact, the whole question of polling through two-way text services could lead to social abuses. If these polls are taken seriously, they could be akin to the old Roman arena thumbs-up, thumbs-down method of decision making. Issues of

public importance might be decided more by emotion than logic and without essential deliberation concerning the long-term consequences.

Another socially related problem of videotext is that it could lead to less social interaction. Instead of going to stores, banks, and ticket agencies where they interface with other people, individuals will interface with their TV set or home computer. The work world may become decentralized and people may be able to work at home utilizing home terminals and phone lines. Various types of accounting and clerical jobs could be farmed out this way and at least some higher level work could be done at home. Although this might bring about loneliness and lack of social interaction for some, it could be a blessing for others. The handicapped could work in this fashion, and people with young children could stay at home but still earn an income. If some actual commuting could be turned into telecommuting, the costs and inconvenience of the daily commute could at least be mitigated.

Because videotext is a largely untried medium, many of the issues surrounding it have yet to be explored. Its overall association with conventional publishing and with society in general will surface in the future if, indeed, videotext becomes an important entity.

# reception technologies

15

## Description

The previous chapters dealt with methods of distributing television programming of either a video or text nature. All of these distribution methods could be received on the home TV set as it now stands. This chapter will deal with technologies that may necessitate changes in the way a picture is perceived on the home TV set, namely high definition TV, digital TV, flat screen TV, large screen TV, and 3-D TV.

### High Definition TV

Improved picture resolution would be the primary outcome of high definition TV (HDTV). Resolution is created according to the number of lines of picture information on a screen. In reality there are never actual lines of video information on the screen. A TV picture is created by rapidly moving dots which are activated by electrons. The electrons hit the phosphorescent screen causing it to glow.

There is, however, the potential for 525 lines of glowing dots on standard American TV sets. European sets have 625 lines. Sets in other parts of the world are about equally divided between 525 and 625 lines. As the electron beam moves across and down these 525 or 625 lines, it causes certain ones to glow with varying degrees of intensity depending on the picture which is to be created. Although only a few dots glow at a time, the eye retains the flowing image and creates a constantly moving picture.

The 525 lines from the top to the bottom of the TV screen create what is referred to as 525 lines of resolution. This is an adequate picture, but a picture with more lines of resolution is sharper and clearer.

High definition TV is a plan to increase the lines of resolution to a number greater than either the American 525 lines or the European 625 lines. Different plans with varying numbers of resolution lines have been proposed, but the one that has the greatest acceptance at present is a 1125 line system developed primarily by Sony and NHK (the Japanese Broadcasting Corporation).

The plan for high definition TV includes changing more than just lines of resolution. Some of the changes proposed involve technical qualities such as gamma, colorimetry, and luminance which are of interest primarily to engineers. One that will be very obvious to consumers is a change in aspect ratio. At present TV sets throughout the world are in a 3 by 4 aspect ratio. This means that the picture is four units wide by three units long. Again, various aspect ratios have been proposed, but the one with the highest acceptance is 3 by 5.33. The reason this number was chosen is that it is very similar to the ratio used for the projection of films in theaters. This would increase the compatibility between film and TV. Research has also found that the human optical system, which is used to looking horizontally, absorbs more with a 3:5 aspect ratio.

A controversial point surrounding HDTV involves fields per second. This refers to how many times the electron beam goes from top to bottom of the screen over a set period of time. The American standard is 60 fields per second and the European standard is 50 fields per second. Finding a standard to suit both systems has not been easy. At one time 80 appeared to be a likely compromise number, but lately a movement is afoot to use 60 fields per second.

Various committees have been meeting over the years to attempt to establish a worldwide standard for HDTV. Engineers have realized the difficulties inherent in having different systems among countries. One of the most obvious is that American programming must be dubbed to a different format before it can be broadcast in Europe. These committees have been wrestling not only with the problem of worldwide standardization but also with the problem of making the new HDTV system compatible with the old systems in all countries so that programming transmitted in HDTV can be received on present TV sets.

Obviously, the older sets will not be capable of showing pictures with 1125 lines of resolution. But some mechanism is needed to allow programming to be seen at both 525 and 625 lines of resolution in 3:4 aspect ratio at both 50 and 60 fields per second at the same time that it is seen in whatever the accepted HDTV format turns out to be.

Another technical problem facing HDTV is the use of a great deal more bandwidth and hence spectrum space than conventional TV. Currently TV stations have a bandwidth of six megahertz. HDTV, as it is proposed in the Japanese plan, would use 30 megahertz or five times the amount of spectrum space.

There is only so much spectrum space available, and it is highly sought after not only for broadcasting but also for ship to shore uses, CP radio, military applications, and a host of other purposes. Carving out space for HDTV will not be easy.

Several proposals have been made. One of them was made by CBS in conjunction with Direct Broadcast Satellite. CBS proposed that the DBS spectrum space be allocated to HDTV, but the FCC rejected the idea. Another idea is to combine several UHF channels for HDTV because UHF is underutilized. Engineers are also working on methods of condensing the bandwidth for HDTV so that less spectrum space is needed.

HDTV has a number of technical problems to be resolved before it can become a reality. However, enthusiasm for the concept abounds primarily because the quality of the image is equivalent to that of 35mm film. Technical compromises will no doubt make it a reality at some point in time.[1]

## Digital TV

The use of a digital television signal encoding process improves image quality and also allows television signals to be transmitted or edited without loss of quality. At present TV signals are encoded using the analog method. Analog encodes each TV line in a manner similar to creating a line graph to show a statistical analysis. In other words, there are no discrete points; all measurements are on a continuum—a line that curves up and down. Digital encodes in a discrete way similar to that of looking at individual numbers in a statistical analysis and writing them down in a set order.

When the analog continuous method is used for taking a TV signal from one place to another (e.g. from one videotape recorder to another), the signal changes its shape slightly and this degrades the picture slightly. This is similar to someone trying to retrace a curve on a graph; the reproduction will not be exactly like the original. When the digital method is used to transport a signal, the signal does not waiver or change. As it travels from one source to another separate bits of information transmit with little loss of definition. This would compare to someone copying the numbers of the statistical study; the numbers would be copied accurately, so the results would be undistorted.

With digital technology many generations of program material can be dubbed with little loss of quality. This would be particularly helpful in the editing process. By the time analog material is transferred from one tape to another four or five

**Figure 15.1.** The analog wave changes; the digital remains constant.

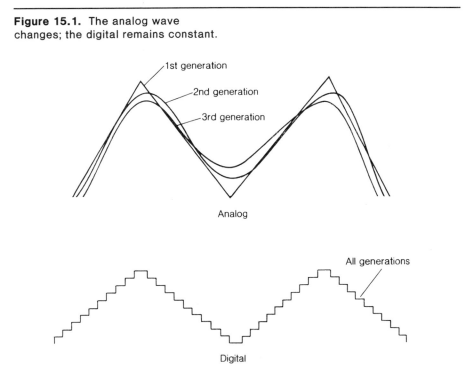

1st generation

2nd generation

3rd generation

Analog

All generations

Digital

times, its quality is noticeably degraded. With the present analog system, pictures lose quality as they are microwaved across the country, a situation that digital technology would improve.

Digital technology can create special effects, both in the production and reception phases. This includes such effects as squeezing, stretching, and flipping the picture. Once home TV sets are digital, viewers will be able to watch several channels on the screen at once or stop a particular picture as it is being transmitted and hold it still on the TV set for a period of time. This is all possible because digital technology can grab at any time any video frame from any video source, change it into digital information, manipulate it in a variety of ways, store it, and retrieve it on command.

Although engineers are trying to invent ways to compress bandwidth, digital TV, like high definition TV, uses more bandwidth.

An international standard has been established for digital TV. This was no easy chore because different countries have different technical parameters surrounding their TV transmission and reception. In designing an international digital standard each country's technical requirements had to be considered.

The two most important elements of digital are fairly technical but they involve sampling frequency (the resolution at which an analog signal is converted to digital) and the ratio of the luminance component of the TV signal to the color components. These elements produce different results within the different national TV technical set-ups. For example, a sampling frequency of 800 might make the best image for British television while a sampling frequency of 900 would be best for Japanese television. After many years of discussions and negotiations, engineers worldwide settled on a sampling frequency of 864 and a luminance to color ratio of 4:2:2.[2]

Now standards are being set for digital tape recorders for characteristics such as width of the tape and nature of the transport system within the cassette.[3]

Digital TV is presently operating, but only with limited applications compared to what it is capable of doing.

## Flat Screen TV

Someday soon television sets may hang on the wall like pictures. This will be due to developments in flat screen TV. Present day TV sets must be deep because they contain a cathode ray tube (CRT). At the back of this tube are electron guns which emit electron beams which hit the phosphorescent TV screen making it glow. Black and white sets contain just one electron gun and color sets contain three—one for reds, one for greens, and one for blues. Physical space is needed for the electrons to travel from the guns to the screen and that is why the TV set has depth.

Two processes have been developed to eliminate the need for this depth. One of these involves flattening the CRT and the other involves liquid crystal displays (LCD).

Sony has been the main purveyor of the first process with its Watchman TV set. This is a pocket-size set that is only an inch and a half thick. Sony accomplished this by putting the electron gun off to the side of the screen and then bending the electron beam at a ninety degree angle. So far they have manufactured only black and white sets using one tube instead of color sets using three.[4]

The other process is marketed primarily by Seiko and Casio and is similar to that used for digital watches. For this process liquid crystals are held between two glass plates. Voltages sent to the liquid crystals make them turn black. In this way a black and white TV picture can be creatd by the presence or absence of impulses.[5]

Both color sets and large flat sets are in development but are not on the market as yet.

**Large Screen TV**

Very closely related to flat screen is large screen TV. Flat screens could be made any size, even as large as theater screens. However, the large screen TVs on the market today use a projection system which throws a picture from inside a modified TV set onto a large screen. Usually there are three projection lenses—one each for red, green, and blue.

Some large screen TVs come in one piece and some in two. Some project from the front and others employ rear screen projection. One common projector is a one piece fold out model with a screen coming from a rectangular cabinet. When in use, a drawer containing a mirror assembly is pulled out and used to throw the picture onto the screen. The two-piece sets have separate screens and projectors. When the projector is not in use, it can often double as a cocktail table. Most rear projection models are one piece and look like extra large TV sets because the projection occurs inside a cabinet onto a translucent screen.

The projected pictures range up to eight feet, and many of the large screen TV systems include provisions for stereo sound. Some also include tailored hook-ups for video games, video cassette recorders, and video disc players. These tailored hook-ups enable these machines to operate through the picture and sound circuits which deliver better quality than the usual method of hooking up to the antenna input.[6]

**Three-D TV**

Three dimensional television is currently possible. Several different methods have been developed for producing it but what is lacking is the economic incentive to market it.

The most commonly used consumer oriented 3-D system is called the anaglyph system. In order to produce a program for this method of 3-D, two different cameras are placed side by side, each taking a slighly different picture. This is to approximate human eyes which are able to perceive depth because they are placed several inches apart. When the program material is played back, two images appear. In order for the viewers to see only one image, they must wear special glasses, each lens filtering out one of the images. For the anaglyph system, one of the images is tinted red and one is green. The viewer wears glasses with one red lens and one green and in that way filters the proper picture to each eye. Movies have been shot using the anaglyph method and have been shown in both theaters and on TV.

A similar method used in movie theaters is the polarization system. Again, two images are shot and the viewer wears glasses so that the images are filtered out because each lens in the glasses is polarized in a different direction. This system has not worked successfully on TV because there is no way to polarize light in two directions on the front of present day TV picture tubes. However, a process is in development which would send the polarized signals to two different TV channels, one for the right eye and one for the left. A viewer would watch both channels on a special set and wear the polarized glasses and in that way the image would appear 3-D.

Another method called the split-image system is used in industrial and military applications. For this the screen is split in the vertical plane and the two images sent to the plane are seen through a hood which contains an elaborate set of lenses and mirrors which permit each eye to focus on one of the pictures. This can only be used for individual viewing because of the need for a hood attached to the TV set.

The direction-selective screen method, also in development, does not require the use of glasses. For this the program is produced using about six different lenses which take the picture in a layered fashion. All of these images are then projected onto the TV screen and approximate the effect of 3-D postcards.

Of course, the ultimate in 3-D would be the holograph technique whereby images would appear in the middle of the living room. This would involve numerous projectors around the ceiling of the room and is not considered to be a practical method of home oriented 3-D.

The anaglyph method has been used for both movie and TV viewing; the polarization system has been used in theaters; the split image system is used industrially; and the direction selective, and holograph methods have been demonstrated experimentally.[7]

## History

Most of the new reception technologies have been in development for over a decade. Like most actors and actresses, they did not achieve instant stardom. Many years of hard work and experimentation went into their development. Their full fledged acceptance by society is often anticlimatic, especially to the engineers and executives who labored through the birth pangs.

### High Definition TV
Development of the 1125 line HDTV system began in Japan in the early 1970s as a joint effort among NHK, Sony, Panasonic, and Ikegami. The latter three

are equipment manufacturers who divided the development of cameras, monitors, video tape recorders, large screen projectors, and other equipment. NHK is the primary Japanese network which tested many of the concepts. Along the way CBS became interested in the project and contributed a bit reduction process which proved crucial to the technological development.

In the mid-1970s, an American company, Compact Video, began work on its less extensive high definition system that used 655 lines of resolution.

Both systems were shown publicly in the U.S. in 1981. CBS was instrumental in arranging for the 1125 line system to be shown first in February at the Society of Motion Picture and Television Engineers (SMPTE) convention in San Francisco. The engineers were very impressed with its quality, noting that a person could stand three feet from the HDTV set and not see any of the lines. After the SMPTE showing, the system was demonstrated in Washington. CBS particularly wanted the Washington demonstration to convince the FCC commissioners that the satellite space for direct broadcast satellite should be reserved solely for high definition TV. It was not successful in its efforts.[8]

Compact Video demonstrated its system to a SMPTE meeting in the fall of 1981. Engineers were invited to the company's Burbank headquarters where they watched high definition pictures being shot in the studio, then beamed to satellite, and shown on monitors. The demonstration was to show the capability of electronic transmission of high definition signals to theaters. Compact Video also transferred a program made on its 655 line system to film and showed it on a large screen to convince the movie engineers that the quality was as good as 35mm film. Reaction to the 655 line system was not as positive as reaction to the 1125 line system, and eventually this system fell by the wayside.[9]

One of the people who saw the CBS demonstrations and became very enthusiastic was the film director Francis Ford Coppola. He became an advocate of using electronic means to shoot movies and produced two short high definition TV works, "Six Shots" and "Double Suicide."

In 1982, the 1125 line system was shown publicly in both Hollywood and New York. Coppola's films were shown, and a page of newspaper print was shown on both regular TV and high definition TV. Once again the reception was outstanding with people comparing the difference between HDTV and regular TV to that of AM radio and FM stereo.[10]

In 1983 SMPTE formed the Advanced Television System Committee (ATSC) to begin looking into the adoption of worldwide HDTV standards. During the course of its deliberations, several other HDTV plans were proposed, the most significant of which was one from RCA for a 750 line system.[11]

After debating and deliberating on many aspects of HDTV, the ATSC decided, in 1985, to recommend the 1125 line system with a 3:5.33 aspect ratio and a 60 field per second timing rate. This recommendation is scheduled to go to the International Radio Consultative Committee for further deliberation.[12]

The 1125 line recommendation should meet with little resistance because it can easily be downconverted to both 525 and 625 line systems. This will make compatibility with present sets easier than it would be with some of the other proposed resolutions. The 3:5.33 aspect ratio will be a change for everyone but is close enough to 3:4 that it will not cause major inconveniences. The 60 field rate continues to be controversial, however, and may encounter resistance from European countries using the 50 field rate.

Meanwhile experimentation in HDTV equipment has been progressing, mainly through the leadership of Sony. That company has developed HDTV cameras, switchers, videotape recorders, and monitors which it has displayed at various conventions to enthusiastic crowds.[13]

Several film makers have experimented with programming concepts and plans are under way to use HDTV with videocassettes, videodiscs, cable TV and other systems that do not use spectrum space. These are ideal modes for HDTV because they can use prototype production and reception equipment without having to worry about the bandwidth problem inherent in other forms of distribution.

High definition TV has had slow but steady progress. No one questions that its high quality can have many applications in both the film and broadcasting fields. The next several years should see continued progress.

## Digital TV

In general, digital technology developed from computers. Its uses in that field demonstrated its accuracy and led to its application elsewhere.

The Society of Motion Picture and Television Engineers was instrumental in the development of digital television. This organization first started investigating the digital experiments in 1972 and acted as watchdog, experimenter, and prodder throughout the development cycle.

From 1972 to 1979, most of the digital work was focused toward encoding the type of TV technology used in the United States, but in 1979 SMPTE recognized the prudence of developing a digital system which could be used worldwide and began incorporating elements which would operate with the other systems.

About the same time digital technology had advanced to the point where it could be used for special effects. A new generation of special effects generators appeared on the market which, through digital technology, were capable of squeeze zooms, and various animation type effects.

In 1981 at the same San Francisco SMPTE convention that featured HDTV, an all digital studio was featured. This was set up at Westinghouse's KPIX(TV) station in San Francisco. Engineers were enthusiastic about digital, too, although the quality of digital is harder to demonstrate than the quality of HDTV because digital's main advantage is that the signal does not degrade with time and use. It does not make pictures look a great deal better, especially to the unintiated viewer.[14]

SMPTE's major chore then became holding international meetings to try to establish a worldwide digital standard. This was finally accomplished early in 1982, but meetings are still held to work out minor standards.

Now digital equipment must be more thoroughly developed and, as with HDTV, experiments must be conducted to try to reduce the bandwidth needed before digital can become commonplace in studios.

## Flat ScreenTV

Flat Screen TV has been under development for over a decade primarily in Japan and the United States by such companies as RCA, Sony, Mitsubishi, Casio, and Seiko. Both the flattened cathode ray tube and liquid crystal display methods have been possible for quite a few years but not at a marketable cost. They have been used in airplanes and space vehicles where weight and space are crucial considerations.

Sony was the first to come out with a consumer oriented small flat screen TV in the form of its pocket sized Watchman. This was introduced in 1982 with a pricetag of $350. Shortly afterward Casio and Seiko introduced pocket and wrist-watch-sized LCD sets at similar prices.[15]

All of the sets presently on the market are small and black and white. Various companies are experimenting with large color sets that can hang on the wall. For the flattened CRT model, this involves finding a place to put three electron guns rather than just one. For the LCD method, this involves finding a material that can create color. Seiko claims to have such a material in the form of new fast-response liquid crystals combined with thin-film transistors.[16]

Large flat screen TV will not be available for some time, but when it does become available it will enable TV sets to be large enough to hang on walls and small enough to be carried in a purse.

## Large Screen TV

Breakthrough developments in large screen TV will come when flat screen TV is perfected. Then any size screen, from the tiniest to the largest, will be able to be manufactured and hung on a wall or placed in a drawer.

In the meantime, large screen projected TV will continue to develop. The first projected TV was introduced in the 1970s by Advent. The early models were not very satisfactory because they required a totally dark room in order to see a decent image and then only people seated directly in front of the screen had a good view. Anyone sitting to the side saw a very washed out image. As the years progressed, the technical quality of projected TV improved. Now large screen TV can be viewed in lighted rooms with people seated at least somewhat to the side.[17]

Because of the size of large screen TV set-ups, their main purchasers have been business and educational institutions who have a need to show video material to large groups of people. The screens are often set up in conference rooms where they are used for meetings and speeches.

Both their size and cost (about $1000) have precluded widespread large screen entry into homes. However, fairly compact rear screen projection models have been developed recently which may make greater inroads into homes.

## Three D TV

Inventions involving 3-D TV date back at least forty years. John Baird, a British TV developer, worked on the process during the 1940s, and in 1944 applied for a U.S. patent for what he termed a stereoscopic television system.

During the 1950s, a brief wave of 3-D movies hit U.S. theaters using both the anaglyph and polarized systems. During the same decade 3-D movies were also shown on Mexican TV.

Then the novelty of 3-D wore thin, and its use for entertainment purposes waned for several decades. 3-D experiments continued at a slow pace and military and industrial uses surfaced. For example, 3-D is used to enable operators to manipulate objects remotely in radioactive environments. Similarly, 3-D is helpful for underwater drilling. Most of these applications involved the viewing hood split image method.

Then in the early 1980s, the profusion of new technologies led each to try new gimmicks to obtain subscribers or owners. Along the way, 3-D was rediscovered. SelecTV broadcast some of the old 1950s 3-D movies on its subscription TV systems in Los Angeles and Milwaukee. Subscribers wishing to see the films could obtain the special red and green glasses at a local Sears. These 3-D screenings were quite successful for a short period of time.

The videocassette field rediscovered 3-D also in the early 1980s and released several old movies on video cassettes which sold well. A number of broadcast stations showed 3-D movies, again with positive results. But by the mid 1980s, the 3-D craze had waned, and there was little interest in the subject.

If some large company decided to underwrite the development of 3-D TV, numerous techniques could be perfected and distributed widely within five years. A small Hollywood company, 3D Video Corporation, has been at the forefront of providing industrial 3-D TV and has been the main company working on perfections of the various 3-D systems, but major companies have yet to express an interest in development of the phenomenon.[18]

## Issues

Many of the issues surrounding high density TV, digital TV, flat screen TV, large screen TV, and 3-D, are technological rather than sociological. Yet all of these technologies do have the potential for changing the way people live.

### High Definition TV

High definition TV's first inroad into the sociological structure may come in the form of changes in the motion picture production business. While broadcasters must fret about bandwidth, film production units could easily utilize high definition equipment to produce and distribute films. High definition cameras, switchers, and videotape recorders, now in prototype, could easily replace film cameras and editing equipment.

The film making business is not likely to take kindly to this, however. Unions will fight the introduction of HDTV because the skills needed are not the skills which presently abound in the movie production business. Negative cutters and film processors would find their employment possibilities shrinking to oblivion.

And yet if electronic techniques can be used to produce technically high quality entertainment, this production will be cheaper and quicker than the present methods and will provide much greater possibilities for special effects. Creativity is creativity, and people who are now excellent film editors should be able to transfer their aesthetic judgment to tape editing, even though they will need to learn to operate new types of equipment.

Films produced using HDTV could be transferred to 35mm film and distributed by current means. However, since no present electronic distribution system exists for film, the film industry could obtain satellite space with wide bandwidth and transmit movies through satellites and earth stations to theaters. Or tapes, rather than films, could be distributed through the conventional film distribution chain and exhibited like large screen TV, only with much higher quality.

High definition's acceptance within the present television arena faces formidable obstacles because all equipment, both studio equipment and home TV sets, will need to be replaced. This will not happen quickly, so signals must be sent in such a manner that they can be received by both present day TV sets and HDTV sets. In this way, HDTV can be accepted gradually, the same way color TV was.

It is doubtful that consumers will buy a new TV set just for high definition. But if high definition is combined with digital technology, flat screen TV, large screen TV, and videotext, then all of the technologies may succeed at once. Consumers are likely to buy new sets if the improved picture can also hang on the wall and if that new set greatly enhances their ability to utilize printed information they retrieve from computer data banks and other sources. High definition could prove to be a greater benefit to teletext and videotext than to conventional video programming because its higher resolution is more crucial to the comprehension of words than of moving images.

Another way that HDTV could receive initial acceptance is through home video—particularly videocassettes. If cassettes could be manufactured that could playback on both conventional TV sets and newly manufactured HDTV sets, people might begin seeing the advantages of high definition and begin replacing their old sets with the new HDTV sets.

Of course, before any of this happens, the international standards for HDTV must be established. This process is occurring but at a slow pace. A number of decision deadlines have been set and missed. Engineers should not make hasty decisions, lest they be wrong, but a little more speed and less preoccupation with minutae could help HDTV become an established form more quickly.

## Digital TV

Digital TV does not pose problems in the sociological area. The fact that people can watch several sports events at the same time may be a minor irritant to sports' widows, but, generally, consumers will not be aware of the inner workings of digital technology.

The major international standards have been developed, and work is underway to reduce bandwidth. The production editing process, whether it be for broadcast, cable TV, videocassettes or discs, or movie theaters, will profit greatly from digital technology.

The major problems involve perfecting digital processes, making it cost effective, and marketing it. But overall, its future seems secure.

## Flat Screen TV

Flat screen TV has a fair number of technical bugs to eradicate, but at some future time it could certainly be used for both home viewing and theater viewing. Combined with digital and high density TV, it could greatly improve television quality.

Video art could receive a huge boost from flat screen TV. When the set hanging on the wall like a picture was not being used for TV viewing or information retrieval, it could become a work of art, displaying a moving image rather than the still image of present art work. Video artists could design both abstract and concrete art which could be sold on cassettes or discs or even distributed over satellite. This could be purchased or leased by individuals wishing to display it in their homes.

The small purse-size sets could keep people constantly in touch with the world around them—perhaps to a greater degree than is beneficial. The image of people walking around town with their eyes glued to their miniature TV sets is not a very positive one. Parents could lose total control of what their children watched if the children were able to carry TV sets with them.

Of course, the children might be using the flat screen TVs simply to gather information. If screens can be carried with school books, children can take them to school, plug into data banks or videodisc lessons, and enhance their learning.

When inexpensive TV sets can be large enough for a wall and small enough for a pocket, many uses will surface for them which are not now envisioned.

## Large Screen TV

Many of the factors involved with flat screen TV also affect large screen TV. Although large screen TVs are now available, they would be greatly improved with high density and digital technologies. Video art may have a market with present projected TV, although art seems to fit the hanging on the wall concept better than a bulky TV set.

If flat screen TV does not develop, large screen TV manufacturers will need to reduce the size of the projection system, as opposed to the size of the screen, if their products are to find widespread acceptance in the home market. Also needed is a reduction in the cost of large screen TV.

## Three-D TV

The main issue facing 3-D TV is whether or not anyone really wants it. It can provide spectacular experiences, but is it worth the cost and inconvenience?

Production is much more expensive with 3-D, if for no other reason than it requires extra cameras and extra processing. A market must exist in order for production companies to undertake the added expense. If 3-D does indeed ever take hold, different methods of writing, producing, and directing will need to be developed to take advantage of the added dimension. This should prove an exciting challenge to the creative community.

On the receiving end, the inconvenience of wearing special glasses or a hood inhibits many people from watching for more time than it takes to satisfy their curiosity. The product must be available in order for people to become interested in 3-D, but, in the chicken and egg tradition, people must be interested enough to buy in order for the product to be produced. With no major company looming on the horizon eager to invest in 3-D TV, it may continue to languish as it has in the past.

Even if it does take hold, 3-D may negatively affect society because it will further increase the dependence of individuals on the media for their thrills. People will be less likely to seek real life experience if they can revert to the "reality" of their 3-D TV set.

High definition TV, digital TV, flat screen TV, large screen TV, and 3-D TV all have great potential for interacting with and improving each other. Their political, social, and technical evolution will be interesting to follow.

# part 4
## conclusion

# implicatons and interrelationships    16

## Interrelationships of Media

Exactly which of the new media and the old media will win the survival of the fittest contest for the public's attention and support is unknown. One scenario could see the three traditional networks, NBC, CBS, and ABC, maintaining the same dominance they've known for years. All the newer media would fall by the wayside and the network-station structure would be the main purveyor of information and entertainment. Another scenario could involve national programming coming from direct broadcast satellite, local programming from low-power TV, and interactive services from AT&T over phone lines. Cable TV could lie dormant, its old-fashioned coaxial cables rotting on the telephone poles. On the other hand, cable might become the dominant medium, knocking off one competitor after another in rapid succession. It may develop such a multitude of channels that it alone will handle national, local, and interactive services. Programs will be repeated so often that videocassette recorders will be relegated to the backs of closets.

The competitive marketplace and the specter of regulation will probably prevent any of these scenarios from becoming reality, but the past few years have seen a multitude of changes within the media structure, and future changes are a predestined fact.

### Movies
The movie exhibition industry has already suffered at the hands of television. From 1945 to 1948, movie attendance averaged 90 million people a week. In 1981, that figure had dropped to about 20 million a week.[1] The drop was actually rather precipitous in the late 1940s and early 1950s when television entered the home, drastically changing the entertainment patterns of the American family.

Movie theaters do not stand to lose so much so quickly from the newer media. Patterns of movie-goers have switched from families to young people. Movies are an escape from home for these young peopele, and they will not be lured into the living room regardless of the program content. Still, the multitude of forms of home delivered pay movies presently available will further diminish movie attendance by adults—perhaps to a point where movie theaters can not afford to survive merely to serve the young.

The movie production business, as opposed to the exhibition business, stands only to gain from the new technologies. The line forms to the rear for all the multitude of entities seeking to distribute motion pictures. The yearly contribution to the movie production companies from pay TV alone is $250 million and rising.[2] The scramble among HBO, Showtime, and The Movie Channel, not to mention STV, MMDS, SMATV, videocassettes, videodiscs, DBS, and even low-power TV, can lead only to fuller coffers for the film production companies. But the companies must lose their myopia in terms of producing films intended for theatrical distribution and lean more toward material that will appeal to the more mature audience ensconced in the living room. If Hollywood can not gear to this increase in quantity and change in quality, other areas of the country and the world stand ready to try their hand.

High definition TV is another element which places the movie industry on the brink of a revolution. A vast threat to those who presently make their living in film production, HDTV has the potential for improving quality, speed, and creativity. Pioneers with stature, other than Francis Ford Coppola, will need to jump on the band wagon in order for HDTV to overcome the inertia of the film community. But the young generation of film makers, weaned on TV, should take to the technology with rapid, firm strides that will make it a primary production method.

Changes will occur within the movie production, distribution, and exhibition industries that will alter the interrelationships within the industry as well as the interrelationships between movie makers and the rest of the entertainment-oriented businesses. Public demand for vicarious thrills will never disappear, so an industry geared toward entertaining the public will remain on sure footing. However, the interplay necessary to create healthy balance sheets within the various factions of the industry will be complicated not only as it always has been by the fickle tastes of the public, but also by the successful and unsuccessful power plays of the various new technologies.

## Newspapers

The first "new" media to affect newspapers was radio. The number of newspapers in existence peaked in 1914 with 2250 individual papers and reached a low-point right after World War II with 1749 papers.[3] This encompassed the period of time when radio was flexing its news gathering muscle. The extra editions of newspapers, designed to bring people the latest news when something truly significant happened, died at the hands of radio's faster paced performance.

Television news affected newspapers in perhaps more sinister ways. The number of daily newspapers has not declined greatly since television's bouncing birth, but the power of newspapers has suffered greatly. The large metropolitan dailies, so influential for so long, are now playing second fiddle to well-scrubbed anchor people. These metropolitan dailies have been steadily decreasing while newspapers in small towns and cities have been on the rise. In 1910, 55% of cities with one paper had at least two; today less then 5% of cities have competing newspapers. What's more, circulation is not keeping up with population increase and young people of the TV generation are not avid newspaper readers.[4]

Now newspapers are faced with even more competition in the form of 24-hour satellite delivered TV newscasts, additional local news coverage by cable local origination and low-power TV, and the looming threat of a burgeoning videotext business. With such a satiation of information possibilities, no one may care about the old-fashioned newspaper, including the advertisers who may be lured with competitive prices to the newer more glamorous media.

With this thought in mind, newspaper publishers are trying to position themselves in the new-fashion world of videotext by becoming the primary purveyors of printed information, whether it be printed ala Gutenberg or sent through an elaborate network of computers and fiber optics. The question remains whether the newspapers have enough power and strength to tackle such competitive giants as AT&T and the cable TV industry, both of whom see videotext as their domain. The spirits of John Peter Zenger, Joseph Pulitzer,and William Randolph Hearst will undoubtedly be rooting for the newspaper factions, but whether the young television-raised generation will provide the leadership needed to strengthen the newspaper position is problematical. They may prefer instead to support the alliance between videotext and video-oriented companies, leaving newspapers to further wither on the vine.

### Broadcasting

The reports of broadcasting's death were premature. During the early 1980s, there were those who were predicting that the broadcasting industry would soon lie dormant, its audience and advertisers attracted by the new suitors.

Indeed, the newer media have eroded some of the network audience, but plenty of folks still tune to ABC, NBC, and CBS, and the network advertising spots sell out quickly.[5]

Networks and stations have made some readjustments in their programming which no doubt contributed to their continued life. Instead of running reruns all summer, they have introduced new programs during that time to keep the audience from turning to the new programming of the cable channels. They air fewer ads during movies in order to be more competitive with the advertising free pay channels. They do not program the quantity of movies that they used to because many people will have already seen the movies on pay channels.

Of all the new media, cable TV and videocassettes are the ones still making broadcasters somewhat nervous. Cable, with its proliferation of programming, does attract audience members who used to watch networks and stations. Videocassettes, with their time-shifting capabilities are an enigma which broadcasters do not entirely understand. Videocassettes are a negative to broadcasters because commercials can be zapped, because many programs taped are never watched, and because people will watch prerecorded videocassettes rather than network programming. On the other hand, network and station programs that are taped and watched might not have been watched otherwise.

Networks seem content with their role as broadcasters. At one time they were rushing headlong into ownership positions within the new media. ABC does own ESPN and CBS is involved in teletext, but for the most part the networks have retreated from attempted infiltration into cable TV, DBS, LPTV, and other newer media forms.

Of course, broadcasters continue to keep their eyes peeled to regulations which might affect them adversely. They are not about to let their guard down on such new media related issues as copyright, must carries, and distant signal imports.

In fact, broadcasters have used the new media to improve their own regulatory lot. They have pointed out to the FCC that the newer media are encumbered with fewer restrictions than the established broadcasting industry. Most of the broadcasting regulations were born because only a limited number of broadcast frequencies were available in the electromagnetic spectrum, and government felt those scarce resources should be carefully controlled so that the people who owned them would serve the public interest, convenience, and necessity. However, as the new media developed, the number of methods by which people could receive television programming became almost limitless. With the vast number of channels possible through cable TV, STV, and low-power TV, not to mention the program possibilities available through videocassettes and videodiscs, the scarcity of resources issue held little water. Broadcasters used this argument to convince federal regulators that their potentially monopolistic power over ideas had evaporated, and that they deserved deregulation because they were but one cog in the information wheel.

Yet, they represent the most powerful cog. The broadcasting industry is well established and well connected. It did not, and will not, simply roll over and die because new technologies present competition.

## Cable TV

Of all the new media, cable TV is the one that has received the most attention. This has been due, in part, to the fact that the predictions regarding the success of cable were extremely rosy for a period of time. Cable TV looked like the legendary pot of gold at the end of the rainbow, but one prediction after another failed.[6]

Many of cable's failures occurred precisely because the predictions had been so rosy. The world will not sit idly by and let one segment of business increase its revenues 95.5 percent in one year. When the appearance created indicates that kind of money will be made for years to come, the predators attack. Obviously the money can be made more than one way.

The pay programming which cable felt it should have to itself began being picked up from the satellites and used for profit making by other entities such as SMATV, MMDS, DBS, and low-power stations. In addition to that, other new technologies such as STV, videocassettes, and videodiscs sold the same movies even though they did not pick them up as a satellite feed.

No sooner had cable established local programming than low-power TV posed the threat of infiltrating that area. Cable began to set itself up as a prime interactive force only to find that the phone companies and newspaper owners had other ideas.

The franchise areas, which were essentially monopolies, proved they could be eroded by SMATV and MMDS. These systems, which did not need to endure lengthy franchise procedures and/or time-consuming cabling processes, could strike at cable quickly. Within a month they could be servicing an area that cable had thought was its own as a wired city.

The cable industry, itself, with its proliferation of programming services, settled in for a period of retrenchment.

Yet, cable has survived. Now that much of the expensive wiring is completed and the hype has dwindled from the programming area, cable may settle in as a stable, economically viable industry. It may become a medium that delivers programming and collects fees from subscribers with little hassle and little overhead. The companies which have managed to survive may soon find their coffers filling.[7]

Through the years cable has established a strong political base. It also has an important ace in the hole, more channels than any of the competition. Coaxial cable, itself, gives cable TV a certain amount of technological superiority. This cable can provide the most complete package of two-way capability, adaptability to new scanning techniques, and flexibility to encompass new programming concepts.

Although times may be looking up for cable, it can not afford to rest. It must keep pace with new technological and programming concepts in order to remain competitive in the fast-paced entertainment world.

## Other New Media

All of the new television technologies are finding that they must compete with each other to eke out their spot in the sun.

MMDS and LPTV both have the potential for reaching approximately the same size audience with the same type of programming. This duplication of service will not be necessary. Teletext and videotext both have the potential for providing the same information. Consumers will not be likely to pay for two methods of textual service. The same movies available on cassettes and discs can be seen over STV or MMDS.

Surely not all the new media will survive when their services are so similar. Politics and marketing will be essential elements of survival. For example, the teletext and videotext services have not yet hit upon a successful marketing strategy. Similarly, video discs, although they have several outstanding features, have not been able to find their niche with the American consumer who will be the one who decides on the long-term viability of each of the new technologies.

An element of cooperation as well as competition is needed within the new technology area in order for survival. Lack of standardization, for example, in such fields as teletext and videocassettes could spell doom in the consumer marketplace. Media such as DBS and LPTV which can combine their traits as a well conceived national-local package should do so in order to strengthen both of them. A marriage of HDTV, digital TV, flat-screen, large screen TV, and videotext would bring all to fruition more quickly. Such cooperation is not frequent among companies or industries which view themselves as competitors, however.

Some of the technologies discussed in this book will become footnotes in history. This phenomenon is not new. A hundred years ago unsuccessful attempts were made to market telegraph services to individual homes. In the 1960s, CBS developed a quasi-film, quasi-tape machine called the EVR which it predicted would replace films, but technical and marketing problems reduced the EVR to oblivion.

Other technologies will survive in a form not even thought of as yet. Radio, during its early development was thought of as a one-to-one medium, and many experimenters were involved in devising methods to make the airwaves private so that messages could be sent confidentially. Only a few farsighted individuals could foresee radio as we know it today.[8]

Still other technologies will thrive in their present form and will grow in stature and power and influence. They must, however, realize that even newer technologies will always be pounding at the door ready to interact with or overtake the old.

## Social Implications

Television, as it presently stands, has an enormous effect on society. The average household TV set is on seven hours a day[9] with the average person watching about three hours of that time.[10] Children spend more time watching TV than they spend in the classroom,[11] and a majority of people consider TV their primary source of news.[12]

The new technologies, in order to succeed financially, will need to add to television use or take time from the established TV use. Either of these is bound to bring significant changes in the social fabric. The new technologies have the potential to change the way we live and the way we are.

### The Individual
The individual will be the main benefactor (or loser) if TV viewing time is increased. If the TV set becomes the primary source of both information and entertainent, each individual will spend more time perched before that set. If the success of pay cable is such that movie theaters disappear; if the success of videotext is such that libraries disappear; if the success of interactive banking is such that banks disappear, then the time originally spent in theaters, libraries and banks will be spent with the home TV set.

This could prove beneficial to the consumer in terms of saved time and convenience. Someone balancing a check book late at night could immediately transfer money to establish a desired balance rather than taking time from work the next day to meet banking hours. Library research would not be limited to library hours. Information that is easily accessible is more likely to be accessed. This could lead to a more informed public.

On the other hand, TV has traditionally been used primarily for entertainment. If added services lead only to more time spent imbibing entertainment, then the individual will become a hedonist sitting in front of the TV switching from movies to sports to video games to 3-D thrills to porno programs. In this way added time becomes wasted time. An individual can vegetate in front of the home TV set exerting energy only to switch from channel to channel or from cable TV to video cassette to video game.

In reality, few people will degenerate to that level, but the potential is there for the lazy (or the demented). Most people will choose from the available programming that which they wish for either information or entertainment. Their range of choices will be much greater than it has been in the past. The variety of programming should be able to fill individual needs and on a time schedule to fit each individual rather than a time frame to suit NBC, CBS, and ABC.

Individuals will be more in control of determining both what they want and what they need. Those wishing to take the easy route will settle themselves in and switch channels until they find something they consider least objectionable fare. Those wishing to use their increased variety toward self improvement will be able to find something to meet their needs.

The new technologies tout interactive services as a primary value to individuals. And, indeed, individuals may be able to profit through interactive shopping and data bank retrieval. But interactive TV is not interactive in a social sense. Although individuals can choose to call up particular information or to progress at their own rate of learning, the interaction that occurs does not imitate the give and take of human interactions and does not develop the individual's social skills. Human beings are basically social animals, and if machine-oriented interaction replaces human interaction, the psychological repercussions may be detrimental.

The individual's privacy may also be threatened by the new technologies. If flat screen TVs are in every room as a window to the world, will these same TVs someday be looking back at us? If the omniscient computer knows our purchase patterns and literary tastes, will it tell on us?

The new technologies may help the individual cope with society in a more effective way, or they may provide such a surplus of information and entertainment choices that individuals find themselves pushing the overload button. The degree to which humans are the master of machines is still in human hands. The decisions made now regarding the power which the new television technologies will encompass will have far-reaching implications.

## Society

The new media are likely to affect varying segments of society in different ways. Rural areas, because they are expensive to cable and inefficient to serve with any low-power short-distance media, may not have access to as many services as large cities. On the other hand, two-way, complete communication systems may decrease the pressures for people to live in clusters in large diverse cities because they will bring to the home many of the services provided by city living. New technologies may be part of a force leading people back to the small towns they have largely abandoned.

The poor, who can not afford many of the new services, may find themselves buried deeper in their ghettos—physically, intellectually, and culturally. The possibility exists that as some of the technologies such as cable and SMATV compete, apartment dwellers may have different media choices than home owners. Minorities may find that their needs are not met or are met in such a narrow manner through specialized programming services that the rest of society is not made aware of those needs.

In a similar vein, the concept of mass media may totally disappear. Network TV is now the only true mass media. Newspapers are fragmenting into smaller and smaller service areas; radio stations program to small audiences of particular music tastes; magazines are broken down to accommodate special interest topics. Only network TV is watched by large numbers of people throughout the country with varying backgrounds and interests. In this way it serves as a national unifying force. If TV, too, becomes fragmented into special interest channels or programming, this unifying force will be lost. Also lost, however, may be the inaneness of many TV programs which must be written and produced to appeal to an overly broad group of people.

One of the special interest types of programs more available now than in the past is "adult" programming, referred to by many as pornography. The abundance of this programming now being produced and consumed may greatly alter the social mindset, leading to increased sexual promiscuity and further demeaning of women both physically and psychologically.

On the other hand, the programming variety provides an opportunity for an increase in cultural and educational programming. Unfortunately, this has not happened. Most of the cultural services that were formed have failed. Education has hardly been tried. With the teaching potential built into videodiscs and interactive services, educational programming should have reached a higher peak than it has. Neither cultural programming nor educational programming have met with the consumer acceptance of "adult" programming. This is not surprising given the preexisting social fabric into which all the new programming has been introduced. But if cultural and educational programming were given a larger role, their acceptance might be greater.

Another form of programming popular with the old media and greatly increased with the new media is sports. With sports programming available 24 hours a day, sports fans can become totally immersed to the exclusion of other interests and responsibilities. The vicarious thrills of sports can create an unreal environment or an environment where winning at all costs becomes the primary goal. Sports viewing, in moderation, however, can be beneficial, especially if it leads to sports participation on the part of the viewer.

The new technologies may serve to bring the family unit closer. If more activities center around the home TV set, the family may interact to a greater degree than at present. On the other hand, if the new technologies lead to even greater TV viewing, family members may each wrap themselves in their own little video shells, interacting with the TV set rather than each other.

An increase in the importance and abilities of the communication industry may greatly affect the transportation industry. If interactive services accommodate working at home and shopping at home, the automobile may lose the stranglehold it now has upon American society. This could help solve both energy and pollution problems, but it would require an enormous change in the social structure that is not likely to happen quickly, if at all.

Communications will undoubtedly assume a more important role in society in the coming years. This in turn will spur new technological developments to improve communication which will then spur even more technological developments. This spiraling relationship between communications and technology will lead to an exciting pace, but one that must be kept in social perspective.

### Economics

Consumers will be the ultimate judges as to which of the new television technologies survive economically. Consumers do not really care about the technologies, themselves. People are interested in watching programs not delivery systems. They want to watch "Rocky," not cable TV or subscription TV or direct broadcast satellite. They want their programming in a convenient, inexpensive form, but they generally have no attachment to any particular delivery method. People also resist change unless they see definite advantages inherent in the change.

For this reason new ideas and technologies must have an economic base that enables them to survive while they are developing popularity. Large companies with deep pockets can provide this economic base, and they are now the main entities involved with the new television technologies.

Just what effect this will have on society and the new technologies is unclear. American society engages in a love-hate relationship with its economic base. Capitalism, free enterprise, corporation, competition, profit, and similar terms are sanctified when compared to similar elements of other economic systems such as communism and socialism. But on the home ground, such words usually have negative connotations. A large corporation making a healthy profit is somehow inherently evil, corrupt, and inflexible, and is obviously gaining at the expense of the "little guy."

The "little guys" of the new technologies have all but vanished. The big corporations are in. Without them, the new technologies could not develop. But unfortunately when big companies come in, they bring along their bad habits.

One of these bad habits is cloning. When one company finds an idea that seems to work, multitudes of others jump on the band wagon. Pay-movies are an excellent example of a program form which has been adopted by virtually everyone.

Sports and hard-R movies have been cloned by various new services. As the big companies come in, they also begin to play the numbers game, impressed only with items which are spoken about in millions and billions. Cable systems which used to be content with a mere 3500 subscribers have been gobbled up by large MSOs whose balance sheets thirst for seven or more figures. Ratings are becoming important to the new media, even to the extent that they are emphasizing the old ideal demographics, women 18 to 45.

A related economic aspect of the new media is advertising. The same ads that have been prevalent on broadcast TV are seeping onto satellite-delivered programs. Although the new technologies have multitudinous opportunities for unique merchandising techniques, the predominant advertising fare is the same insulting talking-stomach-type commercials that have been on broadcast TV for years— the same commercials that have spawned "keep up with the Joneses" materialism and poor eating habits.

Yet a crack in the dike is appearing. Interactive services which can lead to direct point of sale are undertaking innovative methods of presenting product information useful to consumers. Here the approach can be a bit softer because the consumer has already expressed an interest in an item and does not need to be hit over the head with "buy, buy, buy." This type of marketing may reduce the need for the quantity of hard sell sales approaches presently seen in commercials. The narrowcasting concept should also lead to the potential for softened commercials because they can be aimed at people likely to be predisposed to a product rather than at a mass audience.

The new technologies will also affect the operation of businesses, both those involved with the technologies and those in other areas of endeavor. Corporations will be the first to develop wide-scale use of electronic mail. They will also be able to store innumerable records on video discs and will find accessibility to computer data banks a great asset. Teleconferencing may eventually save money and change some of the basic methods by which business is conducted. All of this will both eliminate and create jobs, meaning that people in the workforce will need retraining.

Within the telecommunications industries, the effects of both copyright and piracy will figure into the future. Determination of who pays how much copyright fee to whom will affect the production-distribution cycle, hopefully to insure that sufficient programming will be available to serve the public needs and desires. If piracy gets out of hand, no copyright provisions will help the artistic community, and both the program producer and the consumer will suffer. Neither of these problems are truly under control at present. They represent areas where business needs to concentrate effort.

Telecommunications businesses must constantly assess profitability versus public responsibility. Companies which manufacture and distribute more mundane products such as chairs or paper clips, do not have the social power that companies in the telecommunications field have, allowing them to look at their businesses more in terms of dollars and cents. But television, and the new technologies in particular, must realize that along with gaining profit they must act in a responsible manner to serve the American public.

## Politics

The new technologies have been thrust into international politics. Satellites, in particular, raise the issue of the ownership of outer space. The countries taking the leadership positions in new technology find that other countries are suspicious and dubious about the whole field. International concerns can slow down technological advancement, even though engineers are willing and able to proceed.

Communications leads to a shrinking world which leads to a need for standardized communications so that the various parts of the world can talk to each other easily and economically. The movement toward international standards in such areas as digital TV, HDTV, and teletext is worth the trouble encountered in provincial interests and red tape. However, any type of reliance on the politics of foreign countries is risky to the development of technology as envisioned by the United States.

On the home front, the spirit of deregulation affects the new media. As the possibilities for receiving televised information grow, the need for regulation diminishes. Provisions such as the fairness doctrine seem to have outlived their usefulness if no one programming entity reaches the masses of people. On the other hand, the fate of localism may be in jeopardy if deregulation continues. Only regulation or the threat of it keeps broadcasters serving their local communities. How many programmers, including low-power TV station owners, will want to program material about the local community when the possibility for the more glamorous national programming exists?

The government will need to remain somewhat involved in new media to prevent abuses. Piracy, for example, is something that often needs the help of the FBI, the courts, and the penal system. Copyright needs to be arbitrated by some entity and the government is a logical choice. Technical parameters need to be set, primarily by the FCC, so that mass chaos does not exist in the airwaves or even on wires. Government's watchdog function must also be kept somewhat intact to control the media's power in society.

The multitude of programming possibilities have brought politics closer to the people. Gavel to gavel coverage of the House of Representatives can now be seen on C-Span, and the Senate, if it overcomes its fear of television, may soon be cablecast, too.

If interactive services develop, the public may be given a greater voice in government through two-way polling or voting. This could increase the use of the referendum and initiative and make such political tools more easily accessible to the general public. Petition "signatures" could be gathered through TV, and city council meetings could react to input from citizens in their homes.

The new technologies in their interactions with each other and with society are bound to produce change. Fortunately, many factors within the political, economic, and social systems resist wanton change and demand proof before they accept wholesale turnover. Most skeptical of change is the individual who must visualize something as a positive force in life before accepting it. These factors plus a sense of responsibility on the part of the creators of the new technologies should enable them to serve society in a worthwhile, meaningful way.

# notes

*Chapter 1—Overview*
1. Erik Barnouw, *A Tower in Babel* (New York: Oxford University Press, 1966), p. 66.
2. "State of the Art: Technology," *Broadcasting,* October 10, 1983, pp. 51–74.
3. "Space WARC Primed to Make History," *Broadcasting,* May 13, 1985, p. 82.
4. "Third World Static," *Newsweek,* October 9, 1979, p. 52.
5. "Ratings on the Rebound for Network TV," *Broadcasting,* December 5, 1983, p. 35.
6. "Privacy Law: Race to Pace Technology," *Los Angeles Times,* May 14, 1985, pt. I, p. 1.
7. " '83 Vidpiracy Nears $1 Bil," *Variety,* December 21, 1983. p. 1.
8. "Big Piracy Victory for STV," *Variety,* March 8, 1981, p. 1.
9. "Sex on Pay Television—The Battle Lines Are Formed," *TV Guide,* October 30, 1982, pp. 4–8.

*Chapter 2—Traditional Distribution Processes*
1. Joseph Kerman, *Listen* (New York: Worth Publishers, Inc., 1972), p. 13.
2. The spectrum chart is based on one which appeared with "Third World Static," *Newsweek,* October 8, 1979, p. 52.
3. *The Radio Frequency Spectrum: United States Use and Management* (Washington, D.C.: Office of Telecommunication Policy, 1973.)
4. Milton Kiver, *FM Simplified* (Princeton, New Jersey: Van Nostrand, 1960).
5. "The AM Stereo Question: Motorola or Multisystem," *Broadcasting,* May 7, 1984, p. 95.
6. "Lid's Off FM SCA's," *Broadcasting,* April 11, 1983, p. 35.
7. "Consensus Forms on MTS," *Broadcasting,* January 9, 1984, p. 126.
8. Eugene David, *Television and How It Works* (Englewood Cliffs, New Jersey: Prentice-Hall, 1962), pp. 48–53.
9. William B. Johnston, "The Coming Glut of Phone Lines," *Fortune,* January 7, 1985, pp. 97–100.
10. Sydney W. Head, *Broadcasting in America* (Boston: Houghton Mifflin Co., 1976), pp. 65–66.

*Chapter 3—Satellites*
1. "Today and Tomorrow in Domestic Communications Satellites," *Broadcasting,* May 19, 1980, p. 89.
2. "Satellite Spacing: Multifaceted Debate," *Broadcasting,* March 15, 1982, pp. 136–138.
3. "Communication Satellites: The Birds Are in Full Flight," *Broadcasting,* November 19, 1979, p. 89.

4. "A New Kind of Star Wars: The Battle Over C-Band and Ku-Band Direct," *Broadcasting*, February 18, 1985, p. 44.
5. "Video Programing, Cost Decreases Boost Home Earth Station Use," *Aviation Week and Space Technology*, January 7, 1985, p. 97–103.
6. "Hughes Spacecraft to Serve Cable Television Market," *Aviation Week and Space Technology*, July 15, 1981, p. 54.
7. "FCC to Address Question of Intelsat Competition," *Broadcasting*, December 17, 1984, p. 68.
8. "Communications Satellites: The Birds Are in Full Flight," p. 89.
9. William D. Houser, "Satellite Interconnection," *Television Quarterly*, Fall, 1976, pp. 78–80.
10. Robert N. Wold, "Satellite Distribution Is No Pie in the Sky," *Broadcast Communications*, December, 1979, p. 49.
11. Richard C. Morgan, "Soy Group Gets World TV Report," *The Atlanta Constitution*, August 14, 1979, p. 21.
12. "Demand Exceeds Supply of Westar Transponders," *Broadcasting*, February 4, 1980, p. 23.
13. "Keeping Up with the Satellite Universe," *Broadcasting*, August 17, 1981, p. 32.
14. Robert B. Cooper, Jr., "Home Reception Using Backyard Satellite TV Receivers," *Radio Electronics*, January, 1980, pp. 55–59.
15. "Home Is Where the Dish Is," *Broadcasting*, September 10, 1984, p. 92.
16. "Satellite Delivery Set for Two More Group W Programs," *Variety*, October 30, 1981, p. 1.
17. "Radio Joins Television in Move to Sat'Lite Transmission," *Variety*, April 13, 1981, p. 1.
18. "NBC Cutting the Cord with AT&T," *Broadcasting*, April 8, 1985, p. 156.
19. "Space WARC Primed to Make History," *Broadcasting*, May 13, 1985, pp. 82–90.
20. "Fighting Backyard 'Pirates,' " *Newsweek*, May 27, 1985, p. 83.
21. Lawrence Jordan, "Teleconferencing: Friend or Foe?" *Lodging*, January, 1982, pp. 37–46; and Steven Friedlander, "The World As 'Global Village,' " *Audio-Visual Communications*, May, 1982, pp. 18–26.

*Chapter 4—Computers*

1. John A. Brown and Robert S. Workman, *How a Computer Works* (New York: Arco Publishing Company, 1975), pp. 1–8; and Gary B. Shelly and Thomas J. Cashman, *Introduction to Computers and Data Processing* (Brea, California: Anaheim Publishing Company, 1980), pp. 3.1–4.23.
2. James W. Morrison, *Principles of Data Processing* (New York: Arco Publishing Company, 1979), pp. 36–46.
3. Shelly and Cashman, pp. 5.1–6.27; and Morrison, pp. 51–53.
4. Meg Cox, "Network Systems Grow Fast by Finding Ways to Link Different Kinds of Computers," *The Wall Street Journal*, August 6, 1982, p. 17.
5. Russell A. Stultz, *The Illustrated MS-DOS Wordstar Handbook* (Englewood Cliffs, New Jersey: Prentice-Hall, 1984), pp. 10–16.
6. Fred D'Ignazio, *Messner's Introduction to the Computer* (New York: Julian Messner, 1983), pp. 33–38.
7. Shelly and Cashman, pp. 2.1–2.12
8. T. F. Fry, *Further Computer Appreciation* (London: Newnes-Butterworths, 1977), pp. 3–4.

9. Steven L. Mardell, *Computers and Data Processing* (New York: West Publishing Company, 1979), pp. 31–32.

10. Loy A. Singleton, *Telecommunications in the Information Age* (Cambridge, Massachusetts: Ballinger Publishing Company, 1983), pp. 155–165.

11. Peter W. Bernstein, "Atari and the Video Game Explosion," *Fortune,* July 27, 1981, pp. 40–46.

12. "The Video-Game Shakeout," *Newsweek,* December 20, 1982, p. 75.

13. Kathryn Harris, "Warner Sells Atari to Tramile, Will Report a Loss of $425 Million," *Los Angeles Times,* July 3, 1984, pt. IV, p. 1.

14. Shelly and Cashman, pp. 2.35–2.49.

15. "Privacy Law: Race to Pace Technology," *Los Angeles Times,* May 14, 1985, pt. I, p. 1.

16. Martin O. Holoien, *Computers and Their Societal Impact* (New York: John Wiley and Sons, 1977).

17. "Apple Computer's Bruising Times," *Newsweek,* May 27, 1985, pp. 46–47.

18. M. Lewis, "Computer Crime: Theft in Bits and Bytes," *Nations Business,* February, 1985, pp. 57–58; and "Computer Ethics," *Futurist,* August, 1984, pp. 68–69.

19. "Video Games Zap Harvard," *Newsweek,* June 6, 1983, p. 92.

*Chapter 5—Cable TV*

1. For more information on technical aspects of cable TV, see David L. Willis, "The System Rebuild," *TVC,* December 15, 1980, pp. 154–155; and John P. Taylor, "Not Enough Channel Capacity? Supercable to the Rescue," *Cable Age.* May 18, 1981, pp. 21–32.

2. For more on programming, see "The Cable Connection," *Broadcasting,* May 3, 1982, pp. 37–56; and Peter W. Bernstein, "The Race to Feed Cable TV's Maw," *Fortune,* May 4, 1981, pp. 308–318.

3. "Cable Television is Attracting More Ads: Sharply Focused Programs Are One Lure," *The Wall Street Journal,* March 31, 1981, p. 46.

4. Pat Carson, "Dirty Tricks," *Panorama,* May, 1981, pp. 57–59.

5. David L. Jaffe, "CATV: History and Law," *Educational Broadcasting,* July/August, 1974. pp. 15–16.

6. *Broadcasting/Cable Yearbook, 1981* (Washington, D.C.: Broadcasting Publications, Inc., 1981), p. G–1.

7. For other chronicles of early cable TV, see Mary Alice Mayer Philips, *CATV: A History of Community Antenna Television* (Evanston, Illinois: Northwestern University Press, 1972); and Albert Warren, "What's 26 Years Old and Still Has Growing Pains?" *TV Guide,* November 27, 1976, pp. 4–8.

8. The relationship between cable and government can be found in *Cable Television and the FCC: A Crisis in Media Control* (Philadelphia: Temple University Press, 1973); Steven R. Rivkin, *Cable Television: A Guide to Federal Regulations* (Santa Monica, California: Rand Corporation, 1973; and Martin H. Seiden, *Cable Television USA: An Analysis of Government Policy* (New York: Praeger, 1972).

9. Jaffe, p. 17.

10. *Broadcasting/Cable Yearbook, 1981,* p. G–1.

11. Two sources that deal with early local origination are Ron Merrell, "Origination Compounds Interest with Quality Control," *Video Systems,* November/December, 1975, pp. 15–18, and Sloan Commission on Cable Communications, *On the Cable: The Television of Abundance* (New York: McGraw-Hill, 1972).

12. Two sources dealing with early public access are Richard C. Kletter, *Cable Television: Making Public Access Effective* (Santa Monica, California: Rand Corporation, 1973), and Charles Tate, *Cable Television in the Cities: Community Control, Public Access, and Minority Ownership* (Washington, D.C.: The Urban Institute, 1972).
13. "Cable Survey Shows Growth," *Video Systems* January/February, 1976, p. 6.
14. "Righting Copyright," *Time,* November 1, 1976, p. 92.
15. Margaret B. Carlson, "Where MGM, the NCAA, and Jerry Falwell Fight for Cash," *Fortune,* January 23, 1984, p. 171.
16. "Cumpulsory Licenses Hit Again," *Variety,* April 23, 1981, p. 1.
17. *HBO Landmarks* (New York: Home Box Office, n.d.), p. 1.
18. Sheila Mahony, Nick Demartino, and Robert Stengel, *Keeping Pace with the New Television* (New York: VNU Books International, 1980), p. 61.
19. Don Kowet, "High Hopes for Pay Cable," *TV Guide,* June 10, 1978, pp. 14–18.
20. Sheila Mahony, et. al., *Pace,* p. 131.
21. Ibid., p. 95.
22. "The Dimensions of Cable: 1981," *Channels,* April/May, 1981, p. 88.
23. "CATV Stats Leapin': Pretax Net 45%; Feevee Revenue 85%," *Variety,* December 31, 1980, p. 16.
24. "Cable Revenues Gain Faster Than Profits, Survey Finds," *Broadcasting,* November 30, 1981, p. 52.
25. "The Dimensions of Cable: 1981," p. 88.
26. Ibid., p. 88.
27. "The Top Line: Almost $2 Billion; The Bottom Line: Almost $200 Million," *Broadcasting,* January 5, 1981, p. 75.
28. "Cities Issue Guidelines for Cable Franchising," *Broadcasting,* March 9, 1981, p. 148; and Sheila Mahony, et. al., *Pace,* p. 93.
29. Pat Carson, "Dirty Tricks," *Panorama,* May, 1981, pp. 57–59+; "Fight for TV Franchise in New England Town Elicits Big Bribe Offer," *The Wall Street Journal,* December 22, 1981, p. 1.; and Laurence Bergeen, "Cable Fever," *TV Guide,* June 6, 1981, pp. 23–26.
30. For more on franchizing see "City's Ability to Regulate Cable Television Industry Questioned," *Los Angeles Times,* August 17, 1981, Part II, p. 1.; "Warner Amex Loses Bid to Cablevision for Subscriber TV Franchise in Boston," *The Wall Street Journal,* August 13, 1981, p. 6; "The Gold Rush of 1980," *Broadcasting,* March 31, 1980, pp. 35–56; "Simmering Cable Franchise Issue Comes to A Boil," *Variety,* July 31, 1981, p. 1.; "New York Today Picks Its Cable-TV Winners for Four Boroughs," *The Wall Street Journal,* November 18, 1981, p. 1.; and many other newspaper articles of the late 1970s and early 1980s.
31. "The Top 50 MSO's: As They Are and Will Be," *Broadcasting,* November 30, 1981, p. 37.
32. "Entertainment Analysts Find Cable Mom-Pop Days Are Gone; Big Bucks Rule the Day," *Broadcasting,* June 15, 1981, p. 46.; and "FCC Okays Sale of TPT to Westinghouse, the Largest Communications Merger Ever," *Variety,* July 13, 1981, p. 1.
33. "Cable Television is Attracting More Ads; Sharply Focused Programs Are One Lure," *The Wall Street Journal,* March 31, 1981, p. 46.; "CBS Cable Signs Kraft as Its First Advertiser," *Broadcasting,* August 3, 1981, p. 59; "ARTS Snags GM as Underwriter," *Variety,* September 2, 1981, p. 1.; and "Bristol-Myers to Furnish Series to USA Network," *Broadcasting,* February 23, 1981.

34. "Scribes Back at Typewriters," *Variety*, July 16, 1981, p. 1 and "Pay-TV Settlements Will Hasten Debut of Feevee 1st-Run Prod'n," *Variety*, July 15, 1981, p. 8.
35. Keith Larson, "All You Ever Wanted to Know About Buying an Earth Station," *TVC*, May 15, 1979, pp. 16–19.
36. Speech by Spencer Kaitz, Executive Director of CCTA, at Los Angeles Valley College, May 9, 1981.
37. "Going Once, Twice, Gone on Satcom IV," *Broadcasting*, November 16, 1981, p. 27.
38. "Viacom Becomes Second By-Satellite Pay Cable Network," *Broadcasting*, October 31, 1977, p. 64.
39. "GalaVision Reaches First Anniversary, Free Preview Scheduled for October," *TVC*, October 15, 1980, pp. 33–36; "RCA Pay-Cable Plunge Via 50% Slice of RCTA," *Variety*, May 11, 1981, p. 1; and "Times Mirror Sat'lite Programs Join Feevee Networks on May 1," *Variety*, March 26, 1981, p. 3.
40. "Disney Previews Pay-TV Channel," *Variety*, April 13, 1983, p. 1.
41. "Getty Throws Feevee Bombshell," *Variety*, April 23, 1980; and "Premiere Goes Down Feevee Tube," *Variety*, June 5, 1981, p. 1.
42. Some of the articles which summarize basic cable are: Michael O'Daniel, "Basic-Cable Programming: New Land of Opportunity," *Emmy*, Summer, 1980, pp. 26–30; "The Cable Numbers According to Broadcasting," *Broadcasting*, November 30, 1981; and Peter W. Bernstein, "The Race to Feed Cable TV's Maw," *Fortune*, May 4, 1981, pp. 308–318.
43. For five different views of access programming see: Don Kowet, "They'll Play Bach Backwards, Run for Queen of Holland," *TV Guide*, May 31, 1980, pp. 15–18; Susan Beonarcyzk, "Mid-West Cablers Accent on Access," *Variety*, March 17, 1981, p. 28; "Cable Interconnects: Making Big Ones Out of Little Ones," *Broadcasting*, March 1, 1982, p. 59; Ann M. Morrison, "Part-Time Stars of Cable TV," *Fortune*, November 30, 1981, pp. 181–184; and the entire July, 1981, issue of *Community Television Review* published by the National Federation of Local Cable Programmers.
44. "Week One for Warner's Qube," *Broadcasting*, December 12, 1981, p. 62; and "Second Qube Facility Launched by Warner-Amex Near Cincinnati," *TVC*, October 15, 1980, pp. 58–59.
45. "The Two-Way Tube," *Newsweek*, July 3, 1978, p. 64.
46. "Warner Amex Launches New Retrieval Service in Columbus," *TVC*, February 1, 1981, pp. 68–69.
47. "Times Mirror Sat'lite Programs Join Feevee Networks on May 1," *Variety*, March 26, 1981, p. 3.
48. "Valley Cable Seeks Teen Aud With PPV Shows," *Variety*, May 5, 1983, p. 1.
49. "Home Security Is a Cable TV, Industry Bets," *The Wall Street Journal*, September 15, 1981, p. 25.
50. "Pay Cable TV is Losing Some of Its Sizzle As Viewer Resistance, Disconnects Rise," *The Wall Street Journal*, November 19, 1982, p. 22.
51. "Cable's Lost Promise," *Newsweek*, October 15, 1984, p. 103–105.
52. "Warner Amex's Lowering Expectations," *Broadcasting*, March 19, 1984, p. 37.
53. "Showtime, Movie Channel Unite," *Variety*, September 7, 1983, p. 1.
54. "HBO Takes Best Seat in the House While Rivals Still Wait in Line," *Los Angeles Times*, June 5, 1983, Pt. V, p. 1.
55. "Playboy Ousts Klein," *Broadcasting*, March 19, 1984, p. 54.
56. "NCTA's Programming Marketplace," *Broadcasting*, June 11, 1984, p. 54.

57. "RCA Venture for Cable-TV Programs to End," *The Wall Street Journal,* February 23, 1983, p. 8; and "Spotlight Pay-TV Venture Is Seen Ending With Subscribers Being Shifted to Rivals," *The Wall Street Journal,* September 2, 1983, p. 9.
58. "Programming Setback for HBO," *Variety,* December 19, 1983, p. 1.
59. "CBS, Time Inc. (HBO) and Coca-Cola (Columbia) Join Forces to Prime Pump in Movie Production," *Broadcasting,* December 6, 1982, p. 35.
60. "Nebraska Numbers," *Broadcasting,* October 8, 1984, p. 20.
61. "Antipiracy Movie: Scrambling HBO," *Broadcasting,* February 27, 1982, p. 58.
62. "CBS to Drop Its Cultural Cable-TV Service After Failing to Draw Needed Ad Support," *The Wall Street Journal,* September 14, 1983, p. 7.
63. "Turner the Victor in Cable News Battle," *Broadcasting,* October 17, 1983, p. 27.
64. "Daytime + CHN: Joining Together to Stay Afloat," *Broadcasting,* June 20, 1983, p. 37.
65. "Sorting Through the Fallout of Cable Programing," *Broadcasting,* October 17, 1983, p. 29.
66. "ABC Video Ent. Acquires ESPN," *Variety,* May 1, 1984, p. 1.
67. "Ted Turner's Cable Music Channel Debuts," *Variety,* October 29, 1984, p. 8.
68. "Naming Names," *Broadcasting,* October 8, 1984, p. 20.
69. "Cable TV's Costly Trip to the Big Cities," *Fortune,* April 18, 1983, pp. 82–87.
70. "Warner Amex Cable Cuts Interactive Programming Feed in Six Major Cities," *Variety,* January 19, 1984, p. 1.
71. "RSVP Prez Mohr Outlines Elements for PPV Success," *Variety,* May 18, 1983, p. 2.
72. "Two Major Restraints on Cable Television Are Lifted by the FCC," *The Wall Street Journal,* July 23, 1980, p. 1.
73. "Appeals Court Upholds FCC Repeal of Distant-Signal, Exclusivity Rules," *Broadcasting,* June 22, 1981, p. 32.
74. "Black Tuesday Descends on Cable Industry," *Broadcasting,* May 21, 1983, p. 61.
75. "Free At Last: Cable Gets Its Bill," *Broadcasting,* October 15, 1984, p. 38.
76. "Must-Carry Rules Fall to Court Edict," *Broadcasting,* July 22, 1985, p. 31.
77. "Broadcasters Show Support for NAB-AMST Plan for TV Stereo Must-Carry," *Broadcasting,* November 5, 1984, p. 46.

*Chapter 6—Subscription TV*

1. "STV: Scratching Out Its Place in the New-Video Universe," *Broadcasting,* April 7, 1980, pp. 46–62; and "STV," *Channels,* Field Guide, 1985, p. 42.
2. "48% of ON Viewers Turn on Title Fight," *Variety,* June 14, 1982, p. 4.
3. "Bright Future Seen for STV," *Broadcasting,* November 23, 1981, p. 61.
4. "No Holds Barred for STV," *Broadcasting,* June 21, 1982, p. 23.
5. "After Six Trips to Firing Line, Satellites Finally Put Pay Television on Map," *Variety,* December 9, 1980, p. 1.
6. "Big Piracy Victory for STV," *Variety,* March 8, 1981, p. 1.
7. "FCC Agrees to Erase STV Rules," *Variety,* June 18, 1982, p. 1.
8. "Oak Undertaking Major Expansion of STV Activity," *Variety,* April 11, 1083, p. 1.
9. "Hanging the Crepe for STV," *Broadcasting,* October 15, 1984, p. 46.
10. "ABC to Halt Pay-TV Service Experiment," *Los Angeles Times,* June 14, 1984, Pt. IV, p. 1.
11. "Bloom Is Off STV Rose," *Broadcasting,* September 5, 1983, p. 35.

*Chapter 7—Low-Power TV*

1. "LPTV," *Broadcasting*, February 23, 1981, pp. 39–66.
2. "Another Deluge of LPTV Filings Inundates FCC," *Broadcasting*, February 23, 1981, p. 29.
3. "FCC Whistles Low-Power TV to Standstill," *Broadcasting*, April 13, 1981, p. 28.
4. "Lining Up for Low Power," *Newsweek*, December 21, 1981, p. 76.
5. "LPTV Gets the FCC Go-Ahead," *Broadcasting*, March 8, 1982, p. 35.
6. "Newest TV Stations Are Low in Power and High in Color," *The Wall Street Journal*, October 23, 1984, p. 1; and "Low-Power TV: It Plugs the Gaps," *Los Angeles Times*, December 8, 1984, pt. I, p. 1.

*Chapter 8—Multichannel Multipoint Distribution Systems*

1. "Multichannel MDS," *Channels*, Field Guide, 1985, p. 46.
2. "MDS/Wireless Cable—Boon or Bane," *TVC*, December 15, 1980, pp. 195–201.
3. "FCC Asked to Open Up MDS Band as Challenge to Cable," *Broadcasting*, February 15, 1982.
4. "Appeals Court Supports FCC Regulation of SMATV," *Broadcasting*, December 3, 1984, p. 42.
5. "FCC Reassigns Eight ITFS Channels," *E-ITV*, July, 1983, p. 6.
6. "FCC Gets First MMDS Deluge," *Broadcasting*, October 31, 1983, p. 55.
7. "FCC Releases MMDS Order," *Broadcasting*, February 11, 1985, p. 56.
8. "Wireless Cable Makes Its Washington Debut," *Broadcasting*, December 17, 1984, p. 96.
9. "The Expanding World of Multichannel TV," *Broadcasting*, March 11, 1985, p. 38.
10. "MDS and STV Operators Anxious to Test Their New Anti-Piracy War Club," *Variety*, October 30, 1984, p. 1.

*Chapter 9—Satellite Master Antenna TV (SMATV)*

1. "SMATV," *Channels*, Field Guide, 1985, p. 36.
2. "HBO Will Try Mining SMATV Gold," *Variety*, February 28, 1985, p. 1.
3. "SMATV: The Medium That's Making Cable Nervous," *Broadcasting*, June 21, 1982, pp. 33–46.
4. "Appeals Court Supports FCC Regulation of SMATV," *Broadcasting*, December 3, 1984, p. 42.
5. "HBO Will Try Mining SMATV Gold," p. 1.
6. "Cable Must Pay Rights to Wire Apartments," *Broadcasting*, July 5, 1982, p. 28.
7. "SMATV," p. 36.

*Chapter 10—Direct Broadcast Satellite*

1. "DBS," *Channels*, Field Guide, 1985, p. 48.
2. "Putting Their Money Where Their Applications Are," *Broadcasting*, March 8, 1982, p. 160.
3. "FCC Gets DBS Applications Galore," *Variety*, July 17, 1981, p. 1.
4. "FCC Gives Its Long-Awaited Okay for DBS," *Variety*, June 24, 1982, p. 1.
5. "More Shots Fired at Comsat's DBS," *Broadcasting*, March 30, 1981, p. 54.
6. "Lofty Bid for First DBS System," *Broadcasting*, December 22, 1980, p. 23.
7. "Good News, Bad News in DBS Spacerush," *Broadcasting*, July 20, 1981, p. 23.

8. "CBS-DBS-HDTV: Putting Them All Together for the FCC," *Broadcasting,* July 13, 1981, p. 23.
9. "U.S. Team Back from Geneva, Pleased With Itself and ITU," *Broadcasting,* July 25, 1983, p. 26.
10. Michael Pollan, "How the DBS Kids Stole Comsat's Thunder," *Channels,* July-August, 1983, pp. 39–41.
11. "DBS Makes Its Long-Awaited Bow Today," *Variety,* November 15, 1983, p. 10.
12. "Slow Liftoff For Satellite-To-Home TV," *Fortune,* March 5, 1984, p. 100.
13. "USCI: Plug Pulled," *Broadcasting,* April 8, 1985, p. 44.
14. "Thinning Rank of DBS Pioneers Heads for July 17," *Broadcasting,* July 16, 1984, p. 30.
15. "CBS Will Cut DBS Chord: New Partners Are Elusive," *Variety,* June 29, 1984, p. 1.
16. "Another Nail in the DBS Coffin: Comsat Bows Out," *Broadcasting,* December 3, 1984.

*Chapter 11—Video Cassettes*

1. Ivan Berger, "Video Cassette Recorders: Rising Stars of Home Entertainment," *Popular Electronics,* June, 1980, pp. 51–59; and "VCRs," *Channels,* Field Guide, 1985, pp. 6–8.
2. "VCRs," p. 6.
3. Leonard Shyles, "The Video Tape Recorder: Crown Prince of Home Video Devices," *Feedback,* Winter, 1981, pp. 1–5.
4. Bruce Cook, "High Tech: The New Videocassettes," *Emmy,* Summer, 1980, pp. 40–44.
5. Leonard Shyles, pp. 1–5.
6. "RCA Predicting $1 Bil Retail Year for the VCR Industry," *Variety,* May 13, 1981, p. 1.
7. Bruce Cook, pp. 40–44.
8. "Betamax Decision Overturned," *Variety,* October 20, 1981, p. 1.
9. "High Court to Review the Sony Betamax Case," *Variety,* June 15, 1982, p. 1.
10. Bruce Cook, p. 41.
11. "Betamax Decision Overturned," *Variety,* October 20, 1981, p. 1.
12. "Hollywood Loses to Betamax," *Variety,* January 18, 1984, p. 1.
13. "3d-Quarter VCR U.S. Population Pegged at 13 Mil," *Variety,* October 25, 1984, p. 1.
14. "Kodak Plans To Unveil Its New Qtr-Inch VCR," *Variety,* December 22, 1983, p. 1.
15. "Movie Studios Put More Emphasis on Home Video Pay-TV Markets," *The Wall Street Journal,* May 1, 1984, p. 33.
16. "Closer VCR Tracking," *Broadcasting,* March 4, 1985, p. 7.
17. " '83 Vidpiracy Nears $1 Bil," *Variety,* December 21, 1983, p. 1.
18. "Video Software," *Channels,* Field Guide, 1985, p. 10.

*Chapter 12—Video Discs*

1. "Video Disks," *Broadcasting,* February 2, 1981, pp. 35–42; and "Optical Video Disc," *Channels,* Field Guide, 1985, p. 19.
2. "Space Invaders, Videodiscs, and the 'Bench Connection,' " *Training and Development Journal,* December, 1981, pp. 11–17.
3. "Videodisks Make a Comeback As Instructors and Sales Tools," *The Wall Street Journal,* February 15, 1985, p. 25.

4. "Videodisc Return: You Can See Closely Now," *Los Angeles Times*, January 30, 1985, pt. VI, p. 1.
5. "Magnavision Disk Player Hits the Market," *Broadcasting*, December 18, 1978, p. 77.
6. Neil Hickey, "Take the Videodisc," *TV Guide*, December 26, 1981, pp. 12–14.
7. "Video Disks," p. 36.
8. "RCA Reports Big Success for Its Vidisk Launch," *Variety*, May 1, 1981, p. 1.
9. "RCA Blaming Excessive Optimism for Slow Sales of Videodisc Players," *The Wall Street Journal*, January 26, 1982, p. 25.
10. "RCA Gives Up on Videodisc System," *Los Angeles Times*, April 5, 1984, pt., IV, p. 1.
11. "MCA Still in Vidisk Biz Despite Manufacturing Shifts," *Variety*, February 8, 1982, p. 4.
12. "Laserdisk Players Expand Capabilities," *Variety*, December 27, 1983, p. 7.

*Chapter 13—Teletext*

1. "Teletext," *Channels*, Field Guide, 1985, p. 23.
2. "A Lot of Interest in Teletext at 59th Gathering of NAB," *Variety*, April 14, 1981, p. 8.
3. Ibid., p. 8.
4. Cheryl Rhodes, "The Growth of Videotex in Britain and France," *DataCast*, January/February, 1982, pp. 23–35.
5. Ibid., pp. 23–35.
6. Michael E. Nadeau, "Canada Turns on Telidon," *Desktop Computing*, March, 1982, p. 48.
7. "KNXT and KCET Begin First L.A. Test of Teletext," *Variety*, April 9, 1981, p. 1.
8. "Teletext Test Launched in D.C.," *Broadcasting*, June 29, 1981, p. 65.
9. "Teletext/Videotext Out of Cake in Toronto," *Broadcasting*, May 25, 1981, p. 26.
10. "The Whys and Wherefores of Text Transmission," *Broadcasting*, July 5, 1982, pp. 32–35.
11. "The British Are Coming (With a Teletext Standard)," *Broadcasting*, May 20, 1981, p. 28.
12. "Joint Venture Will Provide Teletext Service During Summer Olympics," *Broadcasting*, January 30, 1984, pp. 72–73.
13. "Teletext Gets Boost with Taft-SSS Venture," *Broadcasting*, February 4, 1985, p. 85.
14. "Taft Introes Its Teletext System in Cincy," *Variety*, July 14, 1983, p. 1.
15. Interview with Craig Udit, Metromedia, October 19, 1984.
16. "Time Closes Its Teletext Door," *Broadcasting*, November 28, 1983, pp. 60–61.
17. Interview with Jane DeLoren, KCBS, October 19, 1984.

*Chapter 14—Videotext*

1. Cheryl Rhodes, "The Growth of Videotext in Britain and France," *DataCast*, January/February, 1982, pp. 23–35.
2. Ibid., pp. 23–35.
3. Maria L. Cioni, "Telidon and Education," *Computer Graphics World*, January, 1982, pp. 30–31.
4. "The Two-Way Tube," *Newsweek*, July 3, 1978, p. 64.
5. "Knight-Rider Launches Viewtron," *Broadcasting*, November 7, 1983, p. 80.
6. Interview with Joanne Tauffel, Times-Mirror, February 8, 1984.

7. Interview with Nancie Hynes, Group W Cable Buena Park, December 12, 1984.
8. "Ma Bell's Big Breakup," *Newsweek,* January 18, 1982, pp. 58–59.
9. "Ma Bell's Mixed Blessing," *Newsweek,* August 23, 1982, p. 52.
10. "Research Firm Predicts Big Market for Videotext Services," *Variety,* August 3, 1981, p. 1.

*Chapter 15—Reception Technologies*

1. "U.S. Industry Adopts NHK Parameters for HDTV," *Broadcasting,* March 25, 1985, p. 68.
2. "One World," *Broadcasting,* March 23, 1981, p. 7.
3. "New Equipment for New Technologies at SMPTE," *Broadcasting,* February 25, 1985, p. 69.
4. Carol S. Goldsmith, "Brave New Television Sets," *American Film,* March, 1983, pp. 25–27.
5. Herbert Shuldiner, "Flat-Screen Color TV," *Popular Science,* November, 1983, pp. 100–102.
6. Gene Bylinsky, "High Tech Hits the TV Set," *Fortune,* April 16, 1984, pp. 70–81.
7. Charles Smith, "The Secrets of Television in Depth," *New Scientist,* January 21, 1982, p. 144.
8. "HDTV: the Look of Tomorrow Today," *Broadcasting,* March 2, 1981, p. 34.
9. "Compact Video's Contribution to HDTV State of the Art," *Broadcasting,* November 2, 1981, p. 25.
10. "HDTV Wows 'em in New York," *Broadcasting,* February 15, 1982, p. 33.
11. "Day of Decision for World HDTV Standard," *Broadcasting,* March 18, 1985, p. 29.
12. "U.S. Industry Adopts NHK Parameter for HDTV," p. 68.
13. "State of the Art," *Broadcasting,* October 8, 1984, p. 58.
14. "One Long Step Closer to Digital TV," *Broadcasting,* February 9, 1981, p. 30.
15. David Lachenbruch, "Television's 50 Years in a Bottle," *Channels,* May/June, 1984, p. 16.
16. Shuldiner, pp. 100–102.
17. David Lachenbruch, "Giant TVS that Pop Up, Tiny TVs that You Can Talk To," *TV Guide,* September 5, 1981, pp. 38–41.
18. "Looking at Television Through Red and Blue Glasses," *Broadcasting,* July 19, 1982, p. 64.

## Chapter 16—Implications and Interrelationships

1. Majority Staff of the Subcommittee on Telecommunications, Consumer Protection and Finance of the Committee on Energy and Commerce of the U.S. House of Representatives, *Telecommunications in Transition: The Status of Competition in the Telecommunications Industry* (Washington, D.C.: U.S. Government Printing Office, 1981), pp. 280–281.
2. Ibid., p. 281.
3. Ray Eldon Hiebert, Donald F. Ungurait, and Thomas W. Bohn, *Mass Media II* (New York: Longman, 1979), p. 225.
4. Don R. Pember, *Mass Media in America* (Chicago: Science Research Associates, 1974), pp. 108–109.
5. "Ad Industry Groups Look Into Crystal Ball," *Broadcasting,* October 29, 1984, p. 46.
6. "Tougher Times for Cable TV," *The New York Times,* July 11, 1982, p. 1.
7. "Market for Cable-TV Systems Booms, As Outlook for the Industry Improves," *The Wall Street Journal,* May 31, 1985, p. 5.
8. Erik Barnouw, *A Tower In Babel* (New York: Oxford University Press, 1966), pp. 28–33.
9. "Glowing Tube, "*Broadcasting,* December 12, 1983, p. 7.
10. The Roper Organization, *Trends in Attitudes Toward Television and Other Media* (New York: Television Information Office, 1983), p. 7.
11. Gerald F. Kline and Philip J. Tichenor, *Current Perspectives in Mass Communications Research* (Beverly Hills, California: Sage, 1972), p. 35.
12. The Roper Organization, p. 5.

# glossary

**access**—to retrieve, as from a computer; a local cable TV channel programmed by community individuals or groups

**addressability**—the ability to send a signal from a central facility that can affect TV sets in outlying facilities, such as homes

**affiliate**—a cable system which receives programming from a cable network or a station which receives programming from a broadcast network

**allocations**—frequency assignments given by the FCC for various communications uses

**alphanumeric**—information sent to a TV set consisting of words, numbers, and graphics

**AM**—amplitude modulation; changing the height of a transmitting wave according to the sound being broadcast

**amplifier**—a circuit, tube, transistor, or other apparatus that draws power from a source other than the input signal and then produces as an output an enlarged reproduction of the essential features of the input

**anaglyph**—a system for creating 3-D pictures for which the viewer must wear special tinted glasses

**analog**—devices or circuits in which the output varies as a continuous function of the input

**antenna**—a wire or set of wires or rods used both to send and to receive radio waves

**Antiope**—the French teletext system

**aspect ratio**—the proportions of the TV picture area; present TV sets have four units of width for every three units of height

**bandwidth**—the number of continuous frequencies within given limits that are allowable for transmission of a given signal

**basic cable**—channels, often advertising supported, for which the subscriber does not pay a large extra fee

**beam**—the flow of electromagnetic radiation in a parallel pattern

**Beta**—a half inch videotape cassette format developed by Sony.

**bicycle**—a method by which the same program is shown on different stations or cable systems at different times because the program is not sent over wires, microwave, or satellite, but rather is mailed, flown, or driven from one station to another.

**bit**—a single unit of storage within a computer that can have either an on or off value.

**booster**—a carrier frequency amplifier which strengthens a signal at one fixed point so it can be retransmitted to another fixed point

**broadband**—ability of a system to operate over a relatively wide range of frequencies—cable TV is sometimes referred to as a broadband communication

**byte**—A string of bits (usually eight) within a computer which represents a particular element of information such as a letter or number.

**cable TV**—a system whereby TV signals are received from various sources such as off-the-air broadcasts and satellite transmissions and sent along coaxial cable to TV receivers

**capacitance**—that property which permits the storage of electrically separated charges when potential differences exist between the conductors

**cassette**—a two-reeled self-contained case for magnetic tape

**cathode ray tube**—a tube in which an electronic beam can be focused to a small cross section on a luminescent screen and can then be varied in position and density to produce what appears to be a moving picture

**Ceefex**—the BBC method of teletext

**character generator**—a device which electronically displays letters or numerals on a TV screen

**chip**—a fingernail-sized element within a computer, usually made of silicon, which stores and processes data

**coaxial cable**—a transmission line in which one conductor completely surrounds the other, creating a cable that is not susceptible to external fields from other sources

**common carrier**—an entity which carries signals for anyone willing to pay

**component**—any portion of a total electronic system such as a transistor on a circuit board or a circuit board in a TV camera or a video disc player in a home video center

**compulsory license**—a copyright fee that is a percentage of earnings rather than a fee for each copyright work used

**computer**—an electronic machine which can perform rapid and complex calculations and also compile, correlate, and select data.

**control track**—the portion of videotape that contains sync information such as horizontal and vertical picture alignment

**converter**—a device for making all the cable TV channels able to appear on the TV screen

**copyright**—an exclusive right to publish, produce, or sell a literary, dramatic, musical, or artistic work

**CPU**—central processing unit; the part of the computer that processes and manipulates data

**cross ownership**—one company owning various media in one market

**data bank**—a large amount of information kept in a computer for access by groups or individuals

**DBS**—direct broadcast satellite; a transmission and reception system through which satellite dishes at people's homes can receive signals from satellites

**decoder**—a device for unscrambling a TV picture signal

**definition**—the sharpness of a picture in terms of its resolution

**demodulate**—to operate on a previously modulated wave in such a way that it will have the same characteristics as the original wave

**deregulation**—the removal of laws and rules that spell out government policies

**descrambler**—a device for unscrambling a TV picture signal

**digital**—devices or circuits in which the output varies in discrete "on-off" steps

**disc**—a device resembling a phonograph record used to play video information

**disconnect**—to cancel a cable or subscription TV service

**dish**—an earth satellite station

**distant signal importation**—a cable system bringing in stations from other parts of the country

**down converter**—a device for changing high frequencies to lower frequency signals

**downlink**—a facility that can receive signals from a satellite

**downstream**—the direction signals are sent from a cable TV system to subscribers

**drop**—cable used to connect a subscriber's set to feeder lines in a cable TV system

**dub**—to make a copy; refers especially to making a copy of a film or tape through electronic processes

**earth station**—the send or receive antenna for satellite transmission

**electromagnetic spectrum**—a continuous frequency range of wave energies including radio waves and light waves

**electron gun**—a structure which produces and controls an electron beam in a TV camera

**electronic publishing**—the providing of material through teletext or videotext

**encode**—to use a code, frequently one composed of binary numbers, to represent individual characters or groups of characters in a message or signal

**equal time**—a rule stemming from Section 315 of the Communications Act which states that TV and radio entities should give the same treatment and opportunity to all political candidates for a specific office

**erase head**—an electromagnet which disturbs the signal previously recorded on a tape prior to recording a new signal

**fairness**—a policy which has evolved from FCC decisions, court cases, and Congressional actions which states that radio and TV stations must present all sides of controversial issues which they discuss

**FCC**—Federal Communications Commission; the administrative governing body which oversees telecommunications on a national level

**feevee**—a term for pay TV

**fiber optics**—the technique of transmitting light or images through glass strands

**field**—one half a complete scanning cycle, or a scanning of all the odd or all the even lines

**flat screen TV**—a form of TV reception being developed that generally does not use a normal picture tube but uses other light emitting materials

**FM**—frequency modulation; placing a sound wave upon a carrier wave in such a way that the number of recurrences is varied

**footprint**—the section of the earth a satellite's signal covers

**format**—a type of videotape recording system used; for example, Beta and VHS are ½″ recording formats

**frame**—one complete scanning cycle of a TV camera or receiver; one individual picture of motion picture film

**franchise**—a special right granted by a government or corporation to operate a facility such as cable TV

**freeze**—immobilization or cessation of an activity, such as a stop in accepting applications for low-power TV

**frequency**—the number of recurrences of a periodic phenomenon, such as a carrier wave, during a set time period, such as a second

**generation**—a rerecording from an original; the first time something is recorded it is first generation, the first rerecording is second generation, the second rerecording is third generation and so on

**gigahertz**—one billion hertz or cycles per second

**ground station**—another name for earth station

**hardware**—equipment used to produce, store, transmit, and receive electronic signals

**head**—a small electromagnet used to read, record, or erase information on magnetic tape

**headend**—the part of a cable TV facility that receives signals from various sources

**helical**—a method of videotape recording whereby video information is placed on the tape at a slant

**hertz**—a frequency unit of one cycle per second, abbreviated Hz

**high definition TV**—television pictures with better resolution than present pictures because they contain over 1000 lines as compared with the present 525 line system

**high speed scanning**—the fast forwarding of tape while still being able to see the picture

**holograph**—a three dimensional image produced by a lensless photographic method that splits a laser beam into two beams

**hub**—part of a cable TV system that receives signals from the headend and sends them by cable to subscribers in its area

**information retrieval**—a system by which material stored in one place can quickly be displayed in another upon request

**informercial**—a cross between a commercial and an informational piece showing various aspects of a product

**interactive**—the capacity for two-way communications

**interconnect**—the linking of different cable systems into mini-networks

**IPS**—inches per second; usually used to indicate tape speed

**ITFS**—Instruction Television Fixed Service; a type of educational transmission operating in the 2500 MHz band which needs a special converter in order to be received

**keypad**—a device used to access material for teletext or videotext

**large screen TV**—projected or flat screen TV shown on a very big viewing surface

**laser**—an acronym for Light Amplification by Simulated Emission of Radiation; a device for transforming light of various frequencies into a very narrow, intense beam

**LED**—light emitting diode; a brightly lighted semi-conductor component

**line-of-sight**—a straight line between a broadcast antenna and a receiver

**local origination**—programs produced about the local community, particularly as it refers to programming created by cable TV systems

**lottery**—a random selection method sometimes used by the FCC to determine who will receive the right to use certain frequencies

**low-power TV**—television stations broadcasting on regular TV station frequencies with much less power than normal TV stations

**M-load**—the tape loading configuration for Beta format videotape recorders

**MATV**—Master Antenna TV; television received in apartment complexes from an antenna on the roof and wires to the various apartments

**MMDS**—Miltichannel Multipoint Distribution Service; an over-the-air commercial service comprised of up to four channels operating in the 2500 megahertz range

**memory**—magnetic information storage which is retrievable

**menu**—a list of material available through teletext or videotext

**microprocessor**—a small solid-state circuit containing memory and sequencing logic which is used to receive and respond to temporary and/or permanent instructions which are in turn able to control other pieces of equipment and their functions

**microwave**—radio waves 1000 MHz and up which can travel fairly long distances

**modem**—an acronym for MOdulator/DEModulator; a device that transforms a typical two-level computer signal into a form suitable for transmission over a telephone wire

**modulate**—to vary the amplitude or frequency of one wave by placing another on it

**monaural**—a single sound source designed for both ears

**monitor**—a device for seeing a TV picture directly from the video output

**MSO**—Multiple System Owner; a company which owns more than one cable TV system

**multiplex**—to transmit two or more messages at the same time on a single channel

**must carry**—stations in the local area that cable TV systems were required to put on their channels

**NABTS**—North American Broadcast Teletext Standard; a system for teletext and video-text proposed by AT&T

**narrowcasting**—putting on program material that appeals to a small segment of the population

**NCTA**—National Cable Television Association; the Washington-based organization which represents the cable industry's interests

**NTSC**—National Television System Committee; the system of television used in the United States

**OFS**—Operations Fixed Service; a microwave service in the 2500 Megahertz range for businesses and similar organizations

**open reel**—a tape recorder for which tape is threaded manually from one reel to another

**Oracle**—the British Independent Broadcasting Authority method of teletext

**page**—a TV screen full of teletext or videotext information

**pay cable**—a system by which subscribers can see commercial free programming for a fee over a cable TV channel

**pay per view**—in subscription TV or pay cable, charging the customer for a particular program watched

**pay TV**—a method by which people pay in order to receive television programming free of commercials; can refer to cable TV, subscription TV, or other forms of distribution

**penetration**—the percent of homes which subscribe to cable TV out of the total homes passed by the cable

**phosphor**—a layer of material on the inner face of the TV tube which floresces when bombarded by electrons

**piracy**—obtaining program material through illegal means and gaining financially from it

**PLP**—Presentation Level Protocol; AT&T's proposed system for teletext and videotext

**polarity**—the positive or negative orientation of a signal; reversed polarity results in a negative picture

**Prestel**—the British system of teletext

**projection TV**—a combination of lenses and mirrors which projects a large television picture onto a screen to obtain a larger display area than that possible through a cathode ray tube

**public access**—programming conceived and produced by members of the public for cable TV channels

**Qube**—an interactive cable television system designed and operated by Warner Amex

**radio frequency**—the portion of the electromagnetic spectrum from about 30 kilohertz to 300,000 kilohertz

**RAM**—random access memory; information in a computer which can be written into and read from and which is capable of being changed

**random access**—being able to call up a particular bit of information that is within a large body of information

**rent-a-citizen**—a practice of cable TV franchising of giving influential citizens stock in the cable TV company

**resolution**—the degree to which fineness of detail can be distinguished in a TV picture

**ROM**—read only memory; information recorded in the memory of the computer when it is manufactured and which can not be changed by the operator

**satellite**—a device orbiting the earth which is capable of both receiving and sending television signals

**scan**—to examine the density of an area point by point and convert that information into an electronic code which can later recreate the density

**scramble**—to change a TV signal electronically so it is not recognizable

**SECAM**—Sequential Couleur a Memorie; the French system for 625 line color television

**Section 315**—the portion of the Communications Act that states that political candidates for the same office must be given equal treatment by television services

**siphon**—a process whereby pay TV systems can drain programming from networks by paying a higher price for it initially

**slant track**—another name for helical recording

**SMATV**—Satellite Master Antenna TV; the placing of satellite dishes on apartment complexes or condominiums so that the residents can receive satellite programming

**SMPTE**—Society of Motion Picture and Television Engineers; a group that gives input regarding various technical standards for television

**software**—program material

**solid state**—circuitry that converts mechanical energy into electrical signals without the use of moving parts

**standardization**—a similarity among various devices of the same type so that they can be interchanged

**stereophonic**—sound reproduction using two channels through two separate speakers to give more feeling of reality

**still frame**—an individual frame of tape being held as one continuous shot

**STV**—subscription TV; scrambled programs broadcast over the air which can be descrambled when a subscriber pays a fee and receives a decoder box

**super 8**—a format of film for which the size of the film used is slightly larger than 8 millimeters

**super station**—a TV station transmitted over satellite to cable TV systems

**supply reel**—the reel on a videotape recorder that contains blank tape or a recorded program prior to it being sent through the VTR

**sync**—various drive pulses, both horizontal and vertical, which maintain the horizontal and vertical scanning process of the video picture as it travels from the camera to its destination

**synchronous satellite**—a satellite that travels in orbit at such a rate that it appears to hang stationary above the earth

**syndicated exclusivity**—the rule which states that a cable TV system must black out a syndicated program from an imported station if that program is playing on a local station

**take-up reel**—the reel onto which videotape is collected during recording or playback

**teleconference**—a meeting in which participants communicate with each other through the use of satellite delivered television signals

**Telematique**—a French national plan designed to bring the technologies of computers and communication to every French household

**Teletel**—the French videotext system

**teletext**—words, numbers, and graphics placed on the vertical blanking interval of a broadcast signal

**Telidon**—the Canadian name for both its teletext and videotext services

**three-D TV**—television screen images which appear to have three dimensions because pictures are taken with several lenses

**tiers**—different amounts of cable TV services for which there are different charges

**toll TV**—an early name for pay TV

**transistor**—a solid-state electronic device that controls current flow without the use of a vacuum

**translator**—a low-power TV transmitter usually used to send a signal into an area of poor reception and now used for low-power TV

**transmitter**—a piece of equipment which generates and amplifies a carrier wave and modulates it with information which can be radiated into space

**transponder**—the part of a satellite that carries a particular program service

**transverse quadraplex**—a type of video recording method which uses 2″ tape and places the signal on the tape vertically

**trap**—a device to prevent a cable TV signal from reaching a particular residence

**tuner**—the portion of a receiver that can select frequencies

**UHF**—ultra high frequency; the area in the electromagnetic spectrum between 300 and 3000 megahertz

**uplink**—a facility that can send a signal to a satellite for further distribution

**upstream**—the direction information is sent from a home to the cable TV facility

**VCR**—video cassette recorder; a device for recording and playing back video information

**VHF**—very high frequency; the area in the electromagnetic spectrum between 30 and 300 megahertz

**VHS**—video home service; the format name for video cassette recorders which use the "M-wrap" configuration

**videocassette**—a magnetic tape in a closed container that can either record or playback video programming when it is inserted into a videocassette tape recorder

**videoconferencing**—another name for teleconferencing

**videodisc**—a device shaped somewhat like a phonograph record that contains video and audio information and that can display this information on a TV screen that is connected to a videodisc player

**video games**—computerized interactive cartridges which can be connected to a TV set or can stand alone in a setting such as an arcade

**videotext**—words, numbers, and graphics which travel through a wired service to the home screen

**viewdata**—highly interactive videotext services

**window**—the time frame for exhibition of a film or video program

**zoom lens**—a lens with a variable focal length which allows a TV camera to frame more or less of a scene without the camera being moved

# index

access, 61–62, 71, 79, 92–93
addressability, 94, 98
"adult" programming. *See* sexually explicit
    programming
Advent, 195
advertising
    on cable TV, 63–64, 76, 81, 88–89, 92
    on new media, 212
    and teletext, 163–164
    and videotext, 174–175, 181
AM, 13–17, 56
American Broadcasting Company (ABC)
    and basic cable, 78
    on cable TV, 70
    and direct broadcast satellite, 133
    dominance of, 6–7
    and early cable TV, 55
    and ESPN, 82, 206
    and low-power TV, 114
    and new media, 205
    and SNC, 82
    and TeleFirst, 104–105
American Express, 76, 77
American Family Theater, 121–122
American Telephone and Telegraph (AT&T)
    launching Telstar, 28
    and newspapers, 205
    as owner of microwave, 20, 29
    as owner of phone wires, 19
    and teletext, 167–168
    and videotext, 174, 180, 182
Ampex, 142
analog, 189
Antiope, 166, 174, 176
Appalachian Community Service Network,
    78
Apple, 45, 174
ARTS, 61, 78, 81
Associated Press, 79
Atanasoff, John V., 42
Atari, 45–46
ATC, 76, 87

Baird, John, 196
bandwidth, 15, 17
basic cable, 61, 77–78, 86
Bell Canada, 177
Bell Laboratories, 44
Berry, Clifford, 42
Beta, 138, 142, 143, 148
Black Entertainment Television (BET), 61,
    87
Blonder Tongue Broadcasting, 104
Bravo, 77, 81
British Broadcasting Corporation (BBC), 33,
    164–165, 175
Bushnell, Nolan, 45–46

Cablecom General, 76
Cable Health Network, 79, 82
Cable Music Channel, 82
Cable News Network (CNN)
    advertising on, 63
    and basic cable, 61, 78
    buying SNC, 82
    and low-power TV, 114
    and MMDS, 118, 121
    and SMATV, 125
cable TV, 55–96
    and broadcasting, 6–7
    and coaxial cable, 19
    and computers, 37
    and direct broadcast TV, 134
    and microwave, 20
    and MMDS, 122
    and networks, 206
    and newer media, 206–207
    and newspapers, 205
    and the poor, 8
    and privacy, 8
    regulation of, 7
    and satellites, 4, 26, 129
    and SMATV, 125–127
    and subscription TV, 106
    and videocassettes, 146–147
    and videotext, 171–173

capacitance electronic disc (CED), 152–153
Capcities, 76
Carson, Johnny, 29
Casio, 195
cathode ray tube (CRT), 18, 39, 41, 190, 195
C-band, 25–26, 129–130
CBN, 78, 82
CBS Cable, 78, 82, 86
Ceefax, 165–167, 174, 175
central processing unit (CPU), 37–39
Channel View Incorporated, 120
Chartwell Communications, 104
Cinemax, 60, 77, 81
"Citizen Kane," 154
coaxial cable, 19, 57
Columbia, 76, 77, 82, 87
Columbia Broadcasting System (CBS)
    and cable TV, 78, 80
    dominance of, 6–7
    and computers, 43
    and early cable TV, 55
    and EVR, 208
    and high definition TV, 132–135, 188,
        193
    and new media, 205
    and teletext, 168, 206
    and Tri-Star, 82, 87
Compact discs, 152
Compact Video, 193
Comp-U-Card, 80
CompuServe, 41, 174
computers, 8, 37–51, 165, 171–173
COMSAT, 28, 33, 131, 133–134
converters, 59
Co-op City, 125
Coppola, Francis Ford, 193, 204
copyright
    and cable TV, 72, 73, 83, 85
    and new media, 212
    and videocassettes, 143, 149
    and videodiscs, 155
    and videotext, 182
Copyright Royalty Tribunal, 73, 83
Cox, 76
cross ownership, 87
C-SPAN, 61, 78, 118, 121

Daytime, 82
deregulation, 83, 105, 206
digital TV, 189–190, 194–195, 197–198,
    208, 213
direct broadcast satellite (DBS), 14, 27, 106,
    129–135, 206–207

Direct Broadcast Satellite Corporation,
    132–133, 188
Disco Network, 79
DiscoVision, 155–157
disconnects, 101
Disney, 31, 46, 60, 77, 143
distant signal importation, 69, 83, 85
Dominion Video Satellite Systems, 132–133

Early Bird, 28
Eckert, J. Presper, 42–43
Eisenhower, Dwight, 43
electromagnetic spectrum, 11–14
ENIAC, 43–44
Entertainment Channel, The, 81, 86
Episcopal Television Network, 78
Eros, 90
ESPN
    and ABC, 82
    advertising on, 63
    and basic cable, 61, 78
    and direct broadcast satellite, 133
    and MMDS, 118
    and satellites, 31
    and SMATV, 125
European Space Agency, 33
EVR, 208

Fairchild, 46
Federal Communications Commission (FCC)
    allocations, 18
    and cable TV, 64, 67–71, 83–84
    and cross-ownership, 87
    and direct broadcast satellite, 131–133
    and HDTV, 188
    history of, 7–8
    and low-power TV, 112–114
    and MMDS, 119–120
    and new media, 206
    and satellites, 28, 33, 76
    and SMATV, 124–125
    and subscription TV, 99, 102, 103–106
    and teletext, 166–167, 169
    and videocassettes, 142
Federal Express, 114
Fedida, Sam, 175
fiber optics, 19
first sale doctrine, 144
flat-screen TV, 190–191, 195, 198, 208
FM, 13–17, 56–57
Focus Broadcasting, 132
Fonda, Jane, 148
FORTRAN, 43

franchising, 65–66, 68, 75, 88
Fraud, 182–183
frequencies, 11–14

Galavision, 77
Gateway, 179
George Mason University, 121–122
Gerard, Emanuel, 46
Getty Oil, 77, 78, 82
Graphics Scanning, 132
Greene, Harold, 180
Griffing, Henry, 103
Group W. *See also* Westinghouse, 179–180

headend, 56–57
Hearst, William Randolph, 205
helical, 137
Hertz, Heinrich, 11
high definition TV (HDTV), 187–188,
    192–194, 196–197
    and direct broadcast satellite, 132–135
    and film making, 204
    and international standards, 213
    and other technologies, 208
Hitachi, 143, 157
Home Box Office (HBO)
    beginnings of, 73–74
    as a cable network, 60
    and low-power TV, 114
    and MMDS, 118–119
    and movies, 204
    and pay-cable, 77
    and satellites, 30–31
    sexual programming on, 90
    slowdowns of, 81
    and SMATV, 126
    and subscription TV rules, 104
    and Tri-Star, 82–87
Home Music Store, 79
home security, 62, 81, 94
Home Team Sports, 121
Home Theater Network, 60
"Hour Magazine," 32
Hubbard Broadcasting, 132–133
hubs, 57
Hughes, 28

IBM, 43–45, 156–157
Ikegami, 192
Independent Broadcasting Authority (IBA),
    165, 175
Instructional Television Fixed Service
    (ITFS), 118–122

Intel Corporation, 44
INTELSAT, 29, 33
interactive TV, 86, 94, 154, 172, 212, 214
interconnects, 93
International Telecommunication Union, 5,
    33

Jewish Network, 78
JVC, 143, 156

Kassar, Raymond, 46
KCET, 164, 166
"King Kong," 154
KLS, 166
Knight-Ridder, 179
KNXT, 166
Kodak, 144
KPIX, 194
KTLA, 102
KTTV, 167
Ku-band, 26, 129

large screen TV, 191, 195, 198, 208
LaserDisc, 157
lasers, 19, 151–153
League of Cities, 84
Learning Channel, The, 61
Lifetime, 61, 82
liquid crystals, 190, 195
local origination, 61
    advertising on, 64
    cutbacks in, 83, 86
    early, 70–71
    growth of, 79
    purpose of, 92
    on STV, 108
lottery, 114, 120
low-power TV, 111–116
    frequency of, 14
    and interference, 4
    and networks, 206
    and other new media, 207
    and satellites, 27
    and subscription TV, 1–6

Magnavox, 46, 143, 155
Master Antenna TV (MATV), 123
Matshshita, 143
Mattel, 46
Mauchly, John W., 42–43
MCA, 46, 77, 155–157
"Merv Griffin Show, The," 32
Microband, 119–120

microwave, 20, 57
Mitsubishi, 195
modem, 41, 171–172
modulation, 14–15
Montgomery Ward, 157
Movie Channel, The
  merging with Showtime, 81–82
  and movies, 204
  and pay-cable, 60, 77
  sexual programming on, 90
  and SMATV, 126
MTV, 61, 78, 82, 118
Multichannel Multipoint Distribution Service
    (MMDS), 117–122
  and direct broadcast TV, 135
  frequency of, 14
  and movies, 204
  and other new media, 207
  and satellites, 27, 129
  and subscription TV, 106
multichannel television, 17
multiple system owners, 68, 70, 76, 81
multiplexing, 16, 18
Multipoint Distribution Service (MDS), 119
must-carries, 69–70, 84–85

Nader, Ralph, 114
narrowcasting, 61
Nashville Network, 78
National Broadcasting Company
  on cable TV, 55, 70
  and direct broadcast satellite, 133
  dominance of, 6–7
  and low-power TV, 114
  and new media, 205
  and teletext, 168
National Christian Network, 78
National Classical Network, 79
National Public Radio, 29
National Aeronautics and Space
    Administration (NASA), 28
newspapers, 178, 204–205
NHK, 187, 192–193
Nickelodeon, 61, 78, 82, 125
Nielsen, 144
North American Broadcast Teletext
    Standard, 167–169, 174

Oak, 104–105, 163
obscenity, 183
Odyssey, 121
ON-TV, 104–105
Operation Fixed Service (OFS), 118–119

Oracle, 165–166, 175
Organization of American States, 33

Panasonic, 168, 192
Paramount, 77, 82, 102–103
Pastore, John, 68
pay cable, 60–61, 63, 77, 82, 86
pay-per-view, 8, 63, 80–81, 83, 100–101,
    108–109
Penthouse, 77
Philips, 155–157
Phonevision, 102–103
Pioneer, 143, 156–157
piracy, 9
  and cable TV, 82, 94–95
  and MMDS, 121
  and new media, 212
  and SMATV, 126
  and subscription TV, 104–105, 109
  and videocassettes, 144, 148–149
  and videodiscs, 155
Playboy, 60, 77, 81, 90
Polaroid, 144
poor, 89, 184, 210
pornography. See also sexually explicit
    programming, 9, 210
Premiere, 77
Presentation Level Protocol, 167, 174
Prestel, 175–176
privacy, 8, 48, 183–184
Private Screenings, 90
Public Broadcasting System, 29
Pulitzer, Joseph, 205

Qube, 80, 177–178, 184

Radio Corporation of America (RCA)
  auctioning transponders, 76
  and direct broadcast satellite, 132
  and The Entertainment Channel, 81
  and flat-screen TV, 195
  and pay-cable, 77
  and satellites, 28, 30–31
  and videocassettes, 143, 156–157
  and videodiscs, 159
  and video games, 46
Radio Shack, 45, 157, 174
Rainbow, 77
random access memory (RAM), 38
ratings, 144, 146, 212
RCTV, 77
read only memory (ROM), 38
Remington-Rand, 43

Reuters, 79
RKO, 103–104
Robert Wold Company, 29
Rockefeller Center, 77, 81
Rogers, 76

Sanyo, 143, 157
Satcom, 76
Satellite Master Antenna TV (SMATV),
    123–127, 129, 204
Satellite News Channel, 78, 82
Satellite Program Network, 78
Satellite Television, 125
Satellite Television Corporation, 131, 133,
    134
satellites, 25–35, 37
scrambling, 95, 97, 109
Sears, 46, 113, 143, 156, 196
Seiko, 195
SelecTV, 104, 121–122, 196
Selectavision, 156–157
sexually explicit programming, 90–91, 98,
    100–101, 107–108, 148
Sharp, 143
Showtime, 60, 77, 81–82, 90, 126, 204
silicon chip, 44
siphoning, 74, 104
Skiatron, 102
SMPTE, 193–194
Sony, 142–143, 187, 191–192, 195
Source, The, 41, 174
Spanish International Network, 61, 63, 79
spectrum, 17–18
Spotlight, 77, 81
stereo, 16–17, 84
Stevenson, Adlai, 43
Storer, 76, 81
subscription TV (STV), 97–109
    and broadcasting, 7
    and computers, 37
    and direct broadcast satellite, 135
    frequency of, 14
    and movies, 204
    and other new media, 207
    and the poor, 8
    rules for, 74
super 8, 141, 148
super stations, 61, 79
Sylvania, 143
syndicated exclusivity, 69, 83, 85

TCI, 76
teleconferencing, 29–30, 34–35

Telediffusion de France, 166
TeleFirst, 105
Telematique, 166
Telemeter, 102–103
Teleprompter, 76, 77, 87
Teletel, 176
teletext, 37, 40, 161–170, 182, 207, 213
Telidon, 166, 174, 177
Telstar, 28
Texaco, 82
three-D TV, 100, 191–192, 196, 199
tiering, 62–63
Time, Inc.
    beginning HBO, 73
    cable magazine, 81
    as cable owner, 76
    and pay-cable, 77
    and SMATV, 126
    and teletext, 168
    and Tri-Star, 82, 87
time-sharing, 41
Times Mirror, 76, 77, 80, 81, 179
Toshiba, 157
TRADIC, 44
Tramiel, Jack, 46–47
transistor, 44
translators, 113
transponders, 26
transverse quadraplex, 137
Tri-Star, 82, 87
Trinity Broadcasting Network (TBN), 78
Turner, Ted
    against networks, 96
    and basic cable, 78
    beginning super stations, 74
    buying SNC, 82
    and ESPN, 82
    and low-power TV, 114
    and teletext, 167
20th Century Fox, 77

UHF
    allocation of, 14
    and early cable TV, 56–57
    and low-power TV, 111
    and MMDS, 117
    spectrum placement of, 18
    and subscription TV, 99, 104, 106
United Artists Productions, 72
United Auto Workers, 114
United Nations, 33
United Press International, 79
United Satellite Communications, 133

UNIVAC, 43
Universal, 143, 157
USA Network, 79
U.S. Satellite Broadcasting Company, 132–133

vacuum tubes, 42
vertical blanking interval, 161–165
very large scale integration (VLSI), 44–45
VH-1, 82
VHF, 14, 17–18, 56–57, 99, 111, 117
VHS, 138, 143, 148
Viacom, 77, 79, 81, 126
video art, 198
video games, 42, 45–47, 51
videocassettes, 137–144
    and broadcasting, 6–7, 205
    and coaxial cable, 19
    and computers, 40
    distribution of, 21
    and HDTV, 197
    and movies, 204
    and other new media, 207
    and the poor, 8
    regulation of, 7
    and subscription TV, 106
    and TeleFirst, 104
    and three-D TV, 196
    and videodiscs, 158–159
videoconferencing, 29–30
videodiscs, 151–159
    and broadcasting, 6
    distribution of, 21
    and movies, 204
    and other new media, 207
    regulation of, 7
    and videotext, 182
videotext, 171–185
    and broadcasting, 6
    on cable TV, 94
    and computers, 4, 37
    and newspapers, 205
    and other new media, 207–208
    and the poor, 8
    and teletext, 170
Viewtron, 179
Von Neumann, John, 43

Warner
    as cable owner, 76
    and basic cable, 78
    and MTV, 82
    and Qube, 80, 83, 177, 184
    and pay-cable, 77
    selling cable systems, 81
    and SMATV, 126
    and video games, 46–47
Watchman, 191, 195
Weaver, Sylvester L. (Pat), 104
Western Union, 28, 132
Westinghouse, 76, 77, 78, 82, 97, 194
WETA, 166
WFTM, 79
WGN, 61, 79
WHCT, 103
window, 99
WKRC, 167
WOR, 61, 79, 102
word processing, 41
World Administrative Radio Conference, 5, 33, 132–133
World Standard Teletext (WST), 167–169, 174
WTBS, 61, 79, 167

Zenger, John Peter, 205
Zenith, 102–104, 143, 157, 167